Jan Hamaekers

Sparse Grids for the Electronic Schrödinger Equation

Jan Hamaekers

Sparse Grids for the Electronic Schrödinger Equation

Construction and Application of Sparse Tensor Product Multiscale Many-Particle Spaces with Finite-Order Weights for Schrödinger's Equation

Südwestdeutscher Verlag für Hochschulschriften

Impressum / Imprint
Bibliografische Information der Deutschen Nationalbibliothek: Die Deutsche Nationalbibliothek verzeichnet diese Publikation in der Deutschen Nationalbibliografie; detaillierte bibliografische Daten sind im Internet über http://dnb.d-nb.de abrufbar.
Alle in diesem Buch genannten Marken und Produktnamen unterliegen warenzeichen-, marken- oder patentrechtlichem Schutz bzw. sind Warenzeichen oder eingetragene Warenzeichen der jeweiligen Inhaber. Die Wiedergabe von Marken, Produktnamen, Gebrauchsnamen, Handelsnamen, Warenbezeichnungen u.s.w. in diesem Werk berechtigt auch ohne besondere Kennzeichnung nicht zu der Annahme, dass solche Namen im Sinne der Warenzeichen- und Markenschutzgesetzgebung als frei zu betrachten wären und daher von jedermann benutzt werden dürften.

Bibliographic information published by the Deutsche Nationalbibliothek: The Deutsche Nationalbibliothek lists this publication in the Deutsche Nationalbibliografie; detailed bibliographic data are available in the Internet at http://dnb.d-nb.de.
Any brand names and product names mentioned in this book are subject to trademark, brand or patent protection and are trademarks or registered trademarks of their respective holders. The use of brand names, product names, common names, trade names, product descriptions etc. even without a particular marking in this work is in no way to be construed to mean that such names may be regarded as unrestricted in respect of trademark and brand protection legislation and could thus be used by anyone.

Verlag / Publisher:
Südwestdeutscher Verlag für Hochschulschriften
ist ein Imprint der / is a trademark of
OmniScriptum GmbH & Co. KG
Heinrich-Böcking-Str. 6-8, 66121 Saarbrücken, Deutschland / Germany
Email: info@svh-verlag.de

Herstellung: siehe letzte Seite /
Printed at: see last page
ISBN: 978-3-8381-1578-8

Zugl. / Approved by: Bonn, University, Diss., 2009

Copyright © 2010 OmniScriptum GmbH & Co. KG
Alle Rechte vorbehalten. / All rights reserved. Saarbrücken 2010

Für Zeynep und Felix Can

Acknowledgments

This book is a revised version of my doctoral thesis with the title "*Tensor Product Multiscale Many-Particle Spaces with Finite-Order Weights for the Electronic Schrödinger Equation*". At this point, I would like to express my gratitude to all people who have contributed to the completion of this thesis. First of all, I would like to thank my adviser Prof. Dr. Michael Griebel for introducing me to this interesting topic as well as for many helpful ideas and numerous inspiring discussions. Additionally, I am very grateful to him for providing an excellent working environment at the Institute for Numerical Simulation of the University of Bonn. Also, I would like to thank the committee member Prof. Dr. Helmut Harbrecht for the interest in my work and acting as the co-referee of this dissertation. Furthermore, I am very grateful to Christian Feuersänger, Margrit Klitz und Ralf Wildenhues for the tedious task of proof-reading. In addition, special thanks go out to Christian Feuersänger, who patiently provided support for his excellent LaTeX PGFPLOTS package. Last but not least, sincere thanks are given to all colleagues at the Institute for Numerical Simulation for the very nice work atmosphere.

Bonn, April 2010   Jan Hamaekers

Contents

1	**Introduction**	**1**
2	**Sobolev Spaces for Many-Particle Functions**	**9**
2.1	Preliminaries	9
	2.1.1 Notation	9
	2.1.2 Fourier transform	10
	2.1.3 Regularity and decay	11
2.2	Many-particle Sobolev spaces	12
2.3	Hyperbolic cross approximation	16
	2.3.1 Multiscale decomposition of general hyperbolic cross spaces	18
	2.3.2 Approximation by general sparse grid spaces	21
2.4	Weighted many-particle spaces	30
	2.4.1 Particle-wise subspace splitting	31
	2.4.2 Many-particle spaces with finite-order weights	35
2.5	Symmetric and antisymmetric many-particle spaces	37
	2.5.1 Symmetric and antisymmetric general sparse grid spaces	41
	2.5.2 Particle-wise splitting of symmetric and antisymmetric spaces	42
2.6	Summary	46
3	**Electronic Schrödinger Equation**	**47**
3.1	Born-Oppenheimer approximation	47
3.2	Antisymmetry principle	48
3.3	Variational formulation	51
3.4	Properties of the solution	52
	3.4.1 Discrete spectrum and exponential bounds	52
	3.4.2 Cusp conditions and regularity results	54
	3.4.3 Decay of weak mixed derivatives	58
3.5	Summary	65
4	**Numerical Methods**	**67**
4.1	Galerkin discretization	67
	4.1.1 Löwdin rules	68
	4.1.2 Parallel eigenvalue solvers	72
4.2	Multiscale Gaussian frame	73
	4.2.1 Gaussians in multiresolution analysis	73
	4.2.2 One-particle multiscale Gaussian frame	76
	4.2.3 Many-particle Gaussian frame	81
4.3	Approximation with finite-order weights	81
	4.3.1 Particle-wise decomposition	82

		4.3.2 Approximation spaces with weights of finite order	84

4.4	Adaptive scheme	86
	4.4.1 A priori choice of the initial approximation space	87
	4.4.2 A posteriori choice of the sequence of approximation spaces	93
4.5	Summary	98

5 Numerical Experiments — 101

5.1	He atom, H_2 and He_2^+ molecules	102
5.2	Lowest fully antisymmetric states of He, H_2 and Li	110
5.3	Li, Be, B and C atoms	115
5.4	LiH, BeH, BH and Li_2 diatomic molecules	121
5.5	Summary	128

6 Conclusions — 131

A Function Spaces — 135

A.1	Hölder spaces	135
A.2	Besov and Triebel-Lizorkin spaces	135

B Wavelets — 137

B.1	Multiresolution analysis and multivariate wavelets	137
B.2	Meyer wavelet family	139

C One- and Two-Particle Operator Integrals — 143

C.1	Integrals involving Gaussians	143
C.2	Integrals involving modulated Gaussians	144

Bibliography — 158

1 Introduction

In the last few decades, the concept of computer simulation has emerged as a third way of practicing science and complements both the classical theoretical and experimental approach. For a computer experiment, a mathematical model is still required, but it does not have to be accessible by analytic means, since the solutions are now obtained approximately by numerical computations. In particular, numerical simulation allows to consider phenomena that may be difficult or impossible to study by experiment in reality because, for example, their time- or length-scale is too small or too large.

As a prominent example, at the scale of molecules, atoms and electrons, all physical systems are in principle described by quantum mechanics. The underlying mathematical model is the Schrödinger equation with an appropriate molecular Hamiltonian. An analytic solution is only possible in a few simple cases and hence, its successful numerical treatment provides an effective predictive tool for a better understanding and new insights about many issues from several research areas like chemistry, biochemistry, molecular physics, solid state physics, material science and nanotechnology.

For most problems in quantum chemistry it is sufficient to consider nuclei as classical point-like particles and to only treat the electrons as quantum-mechanical particles. There, the Born-Oppenheimer approximation of the time-dependent Schrödinger equation leads to a stationary electronic Schrödinger equation with N_{nuc} clamped nuclei $(\mathbf{R}_k, Z_k)$ and N electronic coordinates $(\mathbf{x}_i)$, i.e.

$$H\Psi = E\Psi, \quad H = -\frac{1}{2}\sum_{i=1}^{N}\Delta_i - \sum_{k=1}^{N_{\text{nuc}}}\sum_{i=1}^{N}\frac{Z_k}{|\mathbf{x}_i - \mathbf{R}_k|} + \sum_{i<j}^{N}\frac{1}{|\mathbf{x}_i - \mathbf{x}_j|} + \sum_{k<l}^{N_{\text{nuc}}}\frac{Z_k Z_l}{|\mathbf{R}_k - \mathbf{R}_l|}$$

written in atomic units. The solution of the N-electron problem lives in a $3N$-dimensional space and respects the antisymmetry side-condition imposed by Pauli's principle.

The high dimensionality of this equation turns out to be the main bottleneck for a direct numerical treatment of this problem, since any discretization on e.g. uniform grids with $\mathcal{O}(m)$ points in each direction would involve $M = \mathcal{O}(m^{3N})$ degrees of freedoms which are impossible to store for $N > 1$. Furthermore, only a convergence rate of type

$$\|\Psi - \Psi_M\|_{\mathcal{H}^s} \leq c(N) M^{-r/(3N)} \|\Psi\|_{\mathcal{H}^{s+r}}$$

can be achieved, where $\|\cdot\|_{\mathcal{H}^s}$ is the usual Sobolev norm in $\mathcal{H}^s$, $r+s$ denotes the isotropic smoothness of Ψ and c is a constant which may depend on N but not on M. Here, we encounter the curse of dimensionality [12], i.e. the rate of convergence deteriorates exponentially with the number N of electrons. Therefore, in quantum chemistry further model approximations were developed and implemented which lead to reduced complexity. Here, the standard electronic structure methods are usually called *first principles* or *ab initio* approximation methods, because they are only based on quantum mechanical principles and do not include empirical data. They are widely used to treat problems in

1 Introduction

chemistry, and in 1998 this success was recognized by the award of the Nobel prize to Walter Kohn and John Pople. However, it is not yet clear in a mathematically rigorous way how to systematically improve the approximation error. Therefore, we propose, implement and study a new approach based on the combination of a nonlinear low-rank approximation method and tensor product multiscale bases which provide guaranteed approximation rates. This combination is realized by a special case of a newly formulated general particle-wise subspace splitting of the many-particle space.

In quantum chemistry, typically, model reductions like Hartree-Fock or successive refinements like configuration interaction, coupled cluster, or perturbation theory are employed to achieve tractable approximations to the electronic Schrödinger equation [170]. In addition, density functional theory provides an alternative framework, where the high dimensionality is traded in for a highly nonlinear equation for the one-particle density of the ground state with an unknown but principally exact exchange-correlation part [150]. Moreover, variational and diffusion quantum Monte Carlo methods based on a trial wave function are successfully applied [145].

Hartree-Fock or multi-configuration methods (HF/MCHF) deliver a nonlinear approximation of rank one, or of low rank, respectively, in terms of Slater determinants of one-particle functions, i.e. (neglecting spin)

$$\Psi^{SD} = \frac{1}{\sqrt{N!}} \bigwedge_{p=1}^{N} \phi_p \quad \text{for } \phi_1, \ldots \phi_N \in \mathcal{H}^1(\mathbb{R}^3).$$

Based on the single-particle functions obtained by the low-rank wave functions, there are various methods to improve upon these solutions. A variety of additive methods (configuration interaction), exponential methods (coupled cluster) and perturbative methods (Möller-Plesset theory) are applied successfully and allow to obtain the electron correlation energy of many systems up to high accuracy. Also, low-rank wave functions built from Slater determinants are usually employed to form a Jastrow-Slater trial function in common fixed-node quantum Monte Carlo approaches, e.g. a single-determinant Jastrow-Slater wave function

$$\Psi^{JS} = e^F \Psi^{SD}, \quad F : (\mathbb{R}^3)^N \to \mathbb{R}$$

with a symmetric Jastrow factor in exponential form [53].

For a numerical method, the conventional electronic structure methods mostly expand the one-electron orbitals in an appropriate basis set. Gaussian type orbitals, approximating Slater type orbitals, are the most widely used functions for molecules, because only a few functions are needed for a relatively good approximation and because the resulting integrals can be computed efficiently. Usually, a fixed set of exponents and contraction coefficients is needed for such a basis set. It is commonly obtained from a nonlinear fitting procedure carried out a priori for some sample configurations [47, 189]. Additionally, also wavelet bases are used for the one-particle systems in HF and DFT calculations. For more details see [5, 31, 83, 143, 176, 191] and for a further reading on adaptive wavelet techniques see e.g. [34, 41].

With respect to both the model approximation error and the one-particle discretization error, some existence and convergence theory has been developed [51, 52, 58, 92, 113, 123, 161]. By using basis functions with increasing polynomial degree, the exponential approaches essentially exhibit linear convergence in the number of basis functions [105]. To overcome the limiting effect of the electron-electron cusp on the attainable basis set convergence, explicit two-particle correlation factors are used, e.g. in the framework of the R12 and F12 methods [105] and in the form of Slater type or Gaussian type geminals (STG, GTG) introduced by Boys and Singer [17, 165]. Usually, in quantum Monte Carlo methods, explicit two-particle correlation terms are included in the Jastrow factor of the Jastrow-Slater trial wave function [26, 53]. Note that with the help of a Jastrow factor ansatz, sharp regularity results with generalized Kato cusp conditions were obtained for both the electronic wave function [55] and the electron density [54, 56]. Additionally, Flad, Hackbusch, Schneider, and Kolb [50, 52, 79, 129] used a product ansatz based on a HF solution and a symmetric Jastrow factor together with an adaptive hyperbolic wavelet/sparse grid discretization.

Of course, the specific choice of the basis and/or the trial wave function is crucial for both, the achieved accuracy and the involved overall cost. Especially, the methods that obtain improved correlation energies result in a work complexity of high polynomial degree in the number of degrees of freedom. Exponential complexity in the resulting numerical method for explicit correlation factors is avoided by techniques such as expanding STGs into short sums of GTGs, direct numerical integration [173], introducing a resolution of the identity [114] and the use of an auxiliary basis. With these correlation methods, it is possible to obtain an accuracy suitable for comparison with experiment (chemical accuracy [187]). Additionally, appropriate extrapolation techniques [84] allow to improve upon the convergence rate. The remaining error with respect to experiment stems from model errors resulting from the Born-Oppenheimer approximation and may be reduced further by taking into account leading (scalar) relativistic corrections and (diagonal) Born-Oppenheimer corrections (DBOC). Furthermore, the number of terms necessary for an exact representation is investigated in more detail in [115, 142] and techniques for avoiding the singularities of the Coulomb potentials by e.g. using the transcorrelated Hamiltonian, are studied in [18, 141, 147].

In particular, the standard ab initio quantum chemical methods are used with good success in many practical applications. However, it is not yet clear in a mathematically rigorous way how to systematically improve both the model approximation error and the basis set discretization error simultaneously and efficiently, such that convergence towards the full electronic Schrödinger solution is obtained. Recent reviews on quantitative quantum chemistry and on modern methods for the approximation of electron correlation are given in [86, 175]. Reviews on computational quantum chemistry which also focus on aspects of numerical analysis are given in [25, 120, 121].

For the sake of completeness note that besides the first principle methods, further model approximations based on empirical data in order to treat systems of larger size have been developed. Although these empirical methods are much less accurate, they are also applied with considerable success to various problems in computational chemistry, biochemistry, molecular physics, material science and nanotechnology; see e.g. [16, 65, 66, 68, 70, 74] and the references cited therein.

Tensor product constructions are a common tool for the discretization of multi-dimensional functions. For example, in the multi-configuration Hartree-Fock and the configuration interaction approaches, the many-electron wave function is usually expanded in a sum of Slater determinants with certain side conditions like orthogonality. In [15],

1 Introduction

Beylkin and Mohlenkamp propose an approximation of the many-electron wave function by an unconstrained sum of Slater determinants to hopefully obtain a more efficient expansion. Their approach is based on a special type of tensor product expansion which generalizes expansions that are based on a singular value decomposition in two dimensions to arbitrary dimensions. Here, the non-uniqueness as well as the resulting condition and stability issues are a subject of current research. For more details on (non-unique) low-rank representation of functions and operators in higher dimensions and an algebra for approximate computations in low-rank representations see [13, 14, 80, 81, 110, 154]. In particular, for Coulomb-like singular operators different types of separable approximations can be found in the literature [19, 20, 83]. In practice, these new approaches for low-rank approximation have so far been applied for model problems [14] and the representation of the electron density and the Hartree potential [30].

Sparse grids present a special case of a tensor product expansion for high-dimensional functions. They promise to circumvent the above-mentioned curse of dimensionality of a conventional discretization using full grids, at least to some extent. We envision, up to logarithmic terms, a convergence rate of the type

$$\|\Psi - \Psi_M\|_{\mathcal{H}^s} \leq c(N) M^{-t/3} \|\Psi\|_{\mathcal{H}^{t,s}_{\text{mix}}},$$

where the rate of convergence does no longer exponentially deteriorate with the number N of particles. Now, however, a more restrictive smoothness requirement, namely the boundedness of a certain (t,s)-th mixed derivative is involved.

The principle behind the sparse grid method is the use of a one-dimensional hierarchical multilevel basis. A tensor product construction and a subsequent truncation of the resulting expansion, i.e. the elimination of small coefficients and their corresponding degrees of freedom, then directly lead to sparse grids [24]. Here, depending on the norm and truncation strategy, different variants of sparse grids (regular sparse grids, energy-norm based sparse grids, dimension adaptive sparse grids) can be derived. Furthermore, the sparse grid spaces are related to hyperbolic cross approximation spaces, i.e. spaces of functions with a Fourier transform vanishing outside a hyperboloid region in Fourier space.

In the simplest case, the sparse grid discretization needs only $\mathcal{O}(2^L L^{N-1})$ instead of $\mathcal{O}(2^{LN})$ degrees of freedom with a possibly N-dependent order constant, where 2^L is the number of grid points in one direction for a conventional discretization, and N is the problem dimension. Provided that the considered function possesses bounded mixed second derivatives, the achieved accuracy is of order $\mathcal{O}(2^{-2L} L^{N-1})$ with respect to the $\mathcal{L}^2$-norm. A variant of the sparse grid approach which is based on the energy norm results in a method whose number of degrees of freedom is further reduced. Here, only $\mathcal{O}(2^L)$ degrees of freedom are needed. Nevertheless the achieved accuracy is of the order $\mathcal{O}(2^{-L})$ with respect to the energy norm, which is of the same order as for a conventional discretization on a uniform grid involving $\mathcal{O}(2^{LN})$ degrees of freedom. Thus the L^{N-1}-terms are eliminated and higher-dimensional problems can be treated. However, the constant in the $\mathcal{O}$-notation and the term $\|\cdot\|_{\mathcal{H}^{s+t}_{\text{mix}}}$ may still depend exponentially on N.

Such favourable convergence properties can be achieved for various sparse grid discretization methods in the context of elliptic partial differential and integral equations, quadrature schemes, and financial applications; see [24, 62, 76] and the references cited therein. For the discretization with sparse grids, theoretical investigations can be found in [23, 72, 77]. Moreover, the sparse grid method was amended with adaptive refinement schemes [63] and generalized to wavelet-like multiscale bases. In [72], Griebel and Knapek introduced general sparse grid approximation spaces which in particular include the regular $\mathcal{L}^2$-norm based sparse grid spaces and sparse grid spaces similar to the energy-norm based one as special cases. Furthermore, in [72, 73], optimal possible approximation rates for sparse grid wavelet methods are shown, when used for approximation problems in $\mathcal{H}^s$, where $s \in (-\infty, \infty)$.

When the sparse grid approach is applied to the electronic Schrödinger equation, the respective choice of multilevel basis functions and sparse grid discretization subspaces can also be interpreted as a kind of approximation model. For such a numerical model, a hierarchy of approximate solutions within linear subspaces of the full Hilbert space of functions with given error bounds naturally results. The convergence of the discretization scheme then guarantees convergence of the model hierarchy to Schrödinger's equation. Of course, the question remains how fast this convergence is and how big the costs for the involved discretization scheme and the solution of the resulting discrete eigenvalue problems are. These problems directly relate to the practicality of such a numerical modeling approach.

In [193], Yserentant showed that the solution of the Schrödinger equation lies in certain hyperbolic cross spaces, when the physically relevant antisymmetry of the wave function is taken into account. These spaces are closely related to the smoothness assumption of the sparse grid method. Basically, for the semi-discretization in a regular hyperbolic cross space, an estimate of type

$$\|\Psi - \Psi_K\|_{\mathcal{H}^1} \leq K^{-1/2} \|\Psi\|_{\mathcal{H}^{1/2,1}_{\text{mix}}}$$

could be achieved with respect to the frequency parameter K, where the norm $\|\cdot\|_{\mathcal{H}^{1/2,1}_{\text{mix}}}$ involves bounded mixed derivatives of order $1/2$. Moreover, Yserentant showed in [195] that using regularity and decay properties of the eigenfunctions of the Schrödinger operator, a direct estimation of the approximation error can be achieved which only involves the $\mathcal{L}^2$-norm of the eigenfunction. Here, instead of the regular hyperbolic cross space, a specially scaled hyperbolic cross space is used. Recently, he further generalized his results for hyperbolic cross spaces to the case of spaces spanned by tensor product basis functions which are built from the eigenfunctions of certain one-particle Hamilton operators [196].

A first application of the sparse grid combination technique to the Schrödinger equation was studied in [60]. Furthermore, we implemented and studied sparse grid approaches using Fourier basis functions in [69] and Meyer wavelet bases in [67]. Here, it turned out that, in principle, the sparse grid approach indeed possesses favourable approximation rates and cost complexities for the solution of Schrödinger's equation. However, the computations are limited due to large constants involved in the estimates for the approximation rates and cost complexities.

Nevertheless, in many areas of science, it can be observed that problems of high dimension possess only low *effective* dimensionality, which may be exploited given the right intrinsic coordinates of the problem. The main task here is to find the effective dimension and the correct coordinates. To this end, there is theory from the area of information-based complexity on weighted function spaces and norms with finite-order weights in the setting of reproducing kernel Hilbert spaces [167, 184, 185, 188]. It states

1 Introduction

that function approximation is possible with a complexity which depends on the effective dimension only and not on the total dimension of the problem. For instance, assume that for a high-dimensional function $f : [0,1]^N \to \mathbb{R}$, there exists a dimension-wise decomposition of the form

$$f = \sum_{\mu=1}^{N} \sum_{u \subset \{1,\ldots,N\}, |u|=\mu} f_u$$

with f_u depending on the coordinates $\{x_p\}_{p \in u}$ only. Now, if all terms f_u with $|u| > q$ can be neglected, then it is only the effective dimension, i.e. the lower dimensionality of the q-th order terms, which exponentially enters the work count complexity [64, 112]. Such a type of decomposition is well-known in statistics under the name ANOVA (analysis of variance) [48] and in the field of computational chemistry as a high-dimensional model representation (HDMR) [156]. Recently, an approach based on a dimension-wise decomposition using a bond order dissection and a DFT method is studied in [71].

In this thesis we intend to combine the favorable properties of both, efficient Gaussian type orbitals basis sets, which are applied with good success in conventional electronic structure methods, and tensor product multiscale bases, which provide guaranteed convergence rates and allow for adaptive resolution. To this end, we develop and study a new approach for the treatment of the electronic Schrödinger equation based on a modified adaptive sparse grid technique and a certain particle-wise decomposition with respect to one-particle functions obtained by a nonlinear rank-1 approximation. Here, we employ a multiscale Gaussian frame for the sparse grid spaces and Gaussian type orbitals to represent the rank-1 approximation. In this way, we are able to treat small atoms and molecules with up to six electrons. To our knowledge this is the first time that systems with more than two electrons have been successfully treated by the application of tensor product multiscale bases in the framework of ab initio methods except for HF and DFT methods. So far only model problems and one- and two-electron systems have been dealt with in [50, 60, 67, 69, 129]. Moreover, our approach to combine a nonlinear approximation of low rank and tensor product multiscale bases by a particle-wise decomposition provides the opportunity to be successfully adapted to other high-dimensional problems in the future.

In order to develop our new approach we generalize the sparse grid spaces for high-dimensional functions $f : \mathbb{R}^n \to \mathbb{R}$ introduced in [72] to sparse grid spaces for many-particle functions $f : (\mathbb{R}^D)^N \to \mathbb{R}$ with certain decay conditions for $|\vec{\mathbf{x}}| \to \infty$. For this novel variant of sparse grid spaces we show new estimates for the approximation and complexity orders with respect to the smoothness and decay parameters. We further present a general particle-wise subspace splitting for many-particle Hilbert spaces. This decomposition is based on subspace splittings of the one-particle space and includes, for example, ANOVA-like decomposition spaces and the well-known configuration interaction (CI) spaces as special cases. To our knowledge this is the first time that this connection between these well-known decompositions from different research areas is described. Furthermore, in order to obtain favorably small constants, we also discuss the application of finite-order weights, present a multiscale frame of Gaussian and develop a heuristic h-adaptive refinement scheme.

The remainder of this thesis is organized as follows: In Chapter 2 we introduce and discuss tensor product multiscale many-particle spaces with finite-order weights. To this end, we define many-particle Sobolev spaces, where the membership of a function is characterized by the decay properties of its Fourier transform. We discuss semi-discretization with respect to the frequency parameter by means of general hyperbolic cross spaces and present error estimators. Then, we describe a multiscale decomposition and additional decay properties which lead to general sparse grid spaces and show estimates for the approximation and complexity orders. Furthermore, we introduce a particle-wise subspace splitting which is based on splittings of the one-particle space and discuss weighted many-particle spaces constructed from this splitting. Finally, we deal with the restriction of the newly introduced many-particle spaces to antisymmetric and symmetric subspaces.

In Chapter 3 we review the known regularity and decay properties of the solution of the electronic Schrödinger equation. In particular, we recall the recent results of Yserentant [193, 194, 196] and discuss the resulting approximation orders for appropriate general sparse grid spaces.

We present our new numerical method in Chapter 4. To this end, we shortly discuss the Galerkin discretization for the electronic Schrödinger equation and comment on the setup of the corresponding system matrices as well as on the solution procedure. Furthermore, we introduce a multiscale Gaussian frame and give estimates for its frame constants. Based on this frame and the particle-wise subspace splitting with respect to a nonlinear rank-1 approximation, we present antisymmetric tensor product multiscale many-particle spaces with finite-order weights. Then, we describe our adaptive algorithm to build a sequence of finite-dimensional subspaces for the Galerkin approximation of the electronic Schrödinger equation. We specifically present our scheme for an a priori construction of an appropriate initial subspace for this algorithm.

In Chapter 5 we apply our new approach to small atomic and diatomic systems with up to six electrons, and compare costs, accuracy and convergence rates. Finally, in Chapter 6 we conclude with a summary of our findings and an outlook on interesting future research issues.

Some parts of this thesis are already published in the journal article [69] and the conference proceedings contribution [67].

1 Introduction

2 Sobolev Spaces for Many-Particle Functions

In this thesis we mainly consider high-dimensional many-particle functions. Thus, in this chapter we briefly introduce function spaces, where the according functions describe N particles of dimension D. To this end, after giving some preliminaries, we define standard Sobolev spaces of isotropic smoothness. Then we introduce certain anisotropic Sobolev spaces of dominated mixed smoothness and we discuss the approximation of functions in these Sobolev spaces by band-limited functions, i.e. functions in a general hyperbolic cross space. Furthermore, we consider many-particle weighted spaces, which are related to a particle-wise subspace splitting of the N-particle space. Finally, we discuss the treatment of certain symmetry conditions in the framework of many-particle Sobolev spaces.

2.1 Preliminaries

For a further reading on the Fourier transform and on standard Sobolev spaces we refer to standard textbooks, e.g. the recent monographs [82, 178].

2.1.1 Notation

Let us first introduce some notation conventions used throughout this thesis. We denote the set of positive integers $\{1, 2, 3, \dots\}$ by $\mathbb{N}$ and the set of natural numbers including zero $\{0, 1, 2, \dots\}$ by $\mathbb{N}_0$. We denote the number of particles by N and the dimension of a one-particle space by D. An element $\mathbf{x}$ associated with one particle in $\mathbb{R}^D$ is written in boldface and its d-th component is denoted by $x_{(d)}$, i.e. $\mathbf{x} = (x_{(1)}, \dots, x_{(D)}) \in \mathbb{R}^D$. Vectors in $(\mathbb{R}^D)^N$ are denoted by $\vec{\mathbf{x}} := (\mathbf{x}_1, \dots, \mathbf{x}_N)$ and an element in $\mathbb{N}^N$ e.g. by $\vec{Z} := (Z_1, \dots, Z_N)$. Here, the n-th D-dimensional vector of $\vec{\mathbf{x}} \in (\mathbb{R}^D)^N$ is denoted by $\mathbf{x}_n \in \mathbb{R}^D$. Also, for a vector $\vec{\mathbf{x}} \in (\mathbb{R}^D)^N$ we denote the d-th component of the n-th particle by $x_{n,(d)} \in \mathbb{R}$. Furthermore, if we do not deal with many-particle functions or if there is no number N of particles and no number D of dimensions of a one-particle space specified, we usually denote the number of dimensions by n and write $\vec{x} := (x_1, \dots, x_n) \in \mathbb{R}^n$. Moreover, for a vector $\vec{x} \in \mathbb{R}^n$ we define the Euclidean distance by

$$|\vec{x}|_2 := \sqrt{\sum_{j=1}^n |x_j|^2},$$

the 1-norm by

$$|\vec{x}|_1 := \sum_{j=1}^n |x_j|$$

2 Sobolev Spaces for Many-Particle Functions

and the maximum-norm by
$$|\vec{x}|_\infty := \max_{1 \le j \le n} |x_j|.$$
Also, we denote the cardinal number of a set u by $|u|$ and the (formal) determinant of a matrix A by $|A|$.

For a multi-index $\vec{\alpha} = (\alpha_1, \ldots, \alpha_n) \in \mathbb{N}_0^n$ we define the corresponding linear partial differential operator of order $|\vec{\alpha}|_1$ by
$$D^{\vec{\alpha}} := \frac{\partial^{|\vec{\alpha}|_1}}{\partial_1^{\alpha_1} \cdots \partial_n^{\alpha_n}}.$$
Here, in order to avoid confusion we also explicitly mention the coordinates in the definition of the derivative, i.e. $D_{\vec{x}}^{\vec{\alpha}} = \frac{\partial^{|\vec{\alpha}|_1}}{\partial_{x_1}^{\alpha_1} \cdots \partial_{x_n}^{\alpha_n}}$. Also, for a multi-index $\vec{\alpha} \in \mathbb{N}_0^n$ and a vector $\vec{k} \in \mathbb{R}^N$ we write
$$\vec{k}^{\vec{\alpha}} := k_1^{\alpha_1} \cdots k_n^{\alpha_n}.$$
Furthermore, $a \simeq b$ means that there exist constants C_1 and C_2 independent of any quantities to be estimated such that $C_1 b \le a \le C_2 b$. In this way, $a \lesssim b$ means that a is uniformly bounded by some constant which is a multiple of b. Moreover, the floor function $\lfloor x \rfloor$ is the largest integer less than or equal to $x \in \mathbb{R}$, i.e. $\lfloor x \rfloor = \max\{z \in \mathbb{Z} : z \le x\}$, and the ceiling function $\lceil x \rceil$ is the smallest integer not less than x, i.e. $\lceil x \rceil = \min\{z \in \mathbb{Z} : z \ge x\}$.

Now, let us introduce the Hermitian inner product by
$$\langle f, g \rangle_{\mathcal{L}^2(\Omega)} := \int_\Omega f^*(\vec{x}) g(\vec{x}) \, d\vec{x}$$
and with it the norm
$$\|f\|_{\mathcal{L}^2(\Omega)} := \sqrt{\langle f, f \rangle}_{\mathcal{L}^2(\Omega)} = \sqrt{\int_\Omega |f(\vec{x})|^2 \, d\vec{x}}$$
of the Hilbert space $\mathcal{L}^2(\Omega)$ of square integrable functions. Here, f and g are complex-valued square integrable functions on $\Omega \subset \mathbb{R}^n$. In the following we set $\Omega = \mathbb{R}^n$ and in particular, if the meaning is clear from the context, we write $\mathcal{L}^2$ instead of $\mathcal{L}^2(\mathbb{R}^n)$, $\langle f, g \rangle$ for $\int_{\mathbb{R}^n} f^*(\vec{x}) g(\vec{x}) \, d\vec{x}$ as well as $\|f\|$ for $\|f\|_{\mathcal{L}^2}$. Moreover, as usual we denote the space of square-summable sequences by $\ell^2(\Lambda)$, $\Lambda \subset \mathbb{Z}$, which is a Hilbert space with the Hermitian inner product $\langle \{c_k\}_k, \{d_k\}_k \rangle_{\ell^2(\Lambda)} := \sum_{k \in \Lambda} c_k^* d_k$. The induced norm is given by $\|\{c_k\}_k\|_{\ell^2(\Lambda)} = \sqrt{\sum_{k \in \Lambda} |c_k|^2}$. Note that $\mathcal{L}^2(\Omega)$ and $\ell^2(\Lambda)$ are special cases of the usual Banach spaces $\mathcal{L}^p(\Omega)$ and $\ell^p(\Lambda)$, respectively. Here, for $1 \le p < \infty$ the space of p-integrable functions $\mathcal{L}^p(\Omega)$ is normed with $\|f\|_{\mathcal{L}^p(\Omega)} := (\int_\Omega |f(\vec{x})|^p \, d\vec{x})^{\frac{1}{p}}$ and the space of p-summable sequences $\ell^p(\Lambda)$ is normed with $\|\{c_k\}_k\|_{\ell^p(\Lambda)} = (\sum_{k \in \Lambda} |c_k|^p)^{\frac{1}{p}}$.

2.1.2 Fourier transform

We define the Fourier transform of an absolutely integrable function $f \in \mathcal{L}^1(\mathbb{R}^n)$ by
$$\mathcal{F}[f](\vec{k}) := \left(\frac{1}{\sqrt{2\pi}}\right)^n \int_{\mathbb{R}^n} e^{-i\vec{k}^T \vec{x}} f(\vec{x}) \, d\vec{x}, \tag{2.1}$$

where we also write $\hat{f} := \mathcal{F}[f]$. Note here that $\vec{k}^T\vec{x} = \sum_{j=1}^N k_j x_j$ is equal to the scalar product on $\mathbb{R}^n$. We denote the Schwartz space of all complex-valued rapidly decreasing infinitely differentiable functions in $\mathbb{R}^n$ by

$$\mathcal{S}(\mathbb{R}^n) = \left\{ f \in C^\infty \,:\, \sup_{\vec{x} \in \mathbb{R}^n} |x^{\vec{\beta}} D^{\vec{\alpha}} f| < \infty,\, \vec{\alpha}, \vec{\beta} \in \mathbb{N}_0^n \right\}$$

and its topological dual, the space of all tempered distributions on $\mathbb{R}^n$, by $\mathcal{S}'(\mathbb{R}^n)$. Then, for $f \in \mathcal{S}(\mathbb{R}^n)$, the Fourier transform $\hat{f}$ given by (2.1) is again in $\mathcal{S}(\mathbb{R}^n)$ and in particular the mapping $\mathcal{F}: \mathcal{S}(\mathbb{R}^n) \to \mathcal{S}(\mathbb{R}^n)$ is a linear isomorphism on $\mathcal{S}(\mathbb{R}^n)$. Here, the inverse Fourier transform operator $\mathcal{F}^{-1}$ is given by

$$\mathcal{F}^{-1}[\hat{f}](\vec{x}) = \left(\frac{1}{\sqrt{2\pi}}\right)^n \int_{\mathbb{R}^n} e^{i\vec{x}^T\vec{k}} \hat{f}(\vec{k})\, d\vec{k}. \tag{2.2}$$

Furthermore, it is well-known that both $\mathcal{F}$ and $\mathcal{F}^{-1}$ can be extended to a continuous, linear and bijective operator from $\mathcal{S}'(\mathbb{R}^n)$ onto itself, respectively. Here, the Fourier transform $\mathcal{F}[f]$ of a tempered distribution $f \in \mathcal{S}'(\mathbb{R}^n)$ can be defined by

$$\langle \mathcal{F}[f],\, g \rangle = \langle f,\, \mathcal{F}[g] \rangle \quad \text{for all } g \in \mathcal{S}(\mathbb{R}^n).$$

Note that $\mathcal{S} \subset \mathcal{L}^2 \subset \mathcal{S}'$ and $\mathcal{S}$ is dense in $\mathcal{L}^2$. Moreover, $f \in \mathcal{L}^2(\mathbb{R}^n)$ implies $\hat{f} \in \mathcal{L}^2(\mathbb{R}^n)$ and the Fourier transform operator $\mathcal{F}$ is a unitary isometric automorphism of $\mathcal{L}^2(\mathbb{R}^n)$. Here, for $f, g \in \mathcal{L}^2(\mathbb{R}^n)$ the Fourier transforms $\hat{f}, \hat{g} \in \mathcal{L}^2(\mathbb{R}^n)$ preserve the $\mathcal{L}^2$-inner-product, i.e.

$$\langle f,\, g \rangle_{\mathcal{L}^2} = \langle \hat{f},\, \hat{g} \rangle_{\mathcal{L}^2}. \tag{2.3}$$

Thus, $\mathcal{F}$ is $\mathcal{L}^2$-norm conserving, i.e. $\|f\|_{\mathcal{L}^2} = \|\hat{f}\|_{\mathcal{L}^2}$ for $f \in \mathcal{L}^2(\mathbb{R}^n)$. Also, analogously to the so-called Parseval (or Plancherel) formula (2.3) the relation $\langle f,\, g \rangle = \langle \hat{f},\, \hat{g} \rangle$ holds for $f \in \mathcal{S}'(\mathbb{R}^n)$ and $g \in \mathcal{S}(\mathbb{R}^n)$.

For the tensor product $f \otimes g$ of functions $f, g \in \mathcal{L}^2(\mathbb{R}^n)$ it holds

$$\mathcal{F}[f \otimes g] = \hat{f} \otimes \hat{g},$$

with the usual distributive meaning. Moreover, we define the Fourier convolution of $f, g \in \mathcal{L}^2(\mathbb{R}^n)$ by

$$(f * g)(\vec{x}) := \left(\frac{1}{\sqrt{2\pi}}\right)^n \int_{\mathbb{R}^n} f(\vec{x} - \vec{y}) g(\vec{y})\, d\vec{y}.$$

Here, in particular the relations

$$\mathcal{F}[f * g] = \hat{f}\hat{g} \quad \text{and} \quad \mathcal{F}[fg] = \hat{f} * \hat{g}$$

hold with the usual distributive meaning.

2.1.3 Regularity and decay

We recall some well-known relations concerning the regularity and the decay of a function and its Fourier transform. If for all multi-indices $|\vec{\alpha}|_1 \leq l$ with $l \in \mathbb{N}$ the partial derivatives $D_{\vec{x}}^{\vec{\alpha}} f$ exist and if $\int_{\mathbb{R}^n} |D_{\vec{x}}^{\vec{\alpha}} f(\vec{x})| \, d\vec{x} < \infty$, then the relations $|\mathcal{F}[f](\vec{k})| \lesssim (1+|\vec{k}|_2)^{-l}$ and

$$\mathcal{F}\left[D_{\vec{x}}^{\vec{\alpha}} f\right](\vec{k}) = (i\vec{k})^{\vec{\alpha}} \hat{f}(\vec{k}) \tag{2.4}$$

hold. Especially for the Laplace operator $\Delta = \sum_{j=1}^{N} \frac{\partial^2}{\partial_j^2}$ we obtain the relation

$$\mathcal{F}[-\Delta f] = |\vec{k}|_2^2 \mathcal{F}[f]. \tag{2.5}$$

Moreover, if $\int_{\mathbb{R}^n} |f(\vec{x})|(1+|\vec{x}|_2)^l \, d\vec{x} < \infty$ for $k \in \mathbb{N}$, then its Fourier transform $\hat{f}$ possesses l continuous derivatives and for all partial derivatives with $|\vec{\alpha}|_1 \leq l$ we have the relations

$$D_{\vec{k}}^{\vec{\alpha}} \hat{f}(\vec{k}) = (-i)^{|\vec{\alpha}|_1} \mathcal{F}[\vec{x}^{\vec{\alpha}} f](\vec{k}) \quad \text{and} \quad |D_{\vec{k}}^{\vec{\alpha}} \hat{f}(\vec{k})| \lesssim \int_{\mathbb{R}^n} |\vec{x}|_2^{|\vec{\alpha}|_1} |f(\vec{x})| \, d\vec{x}.$$

We say that a tempered distribution $f \in \mathcal{S}'$ possesses a weak partial derivative $D^{\vec{\alpha}} f$ if

$$\left\langle \phi, D^{\vec{\alpha}} f \right\rangle = (-1)^{|\vec{\alpha}|_1} \left\langle D^{\vec{\alpha}} \phi, f \right\rangle$$

for all test functions $\phi \in \mathcal{S}$.[1]

Let us now consider the case of $f \in \mathcal{L}^2(\mathbb{R}^n)$. First, if and only if $\int_{\mathbb{R}^n} |\vec{k}|_2^{2l} |\hat{f}|^2 \, d\vec{k} < \infty$, the function f has weak partial derivatives $D_{\vec{x}}^{\vec{\alpha}} f$ for all $|\vec{\alpha}|_1 \leq l$. In particular, relation (2.4) then has the usual distributive meaning. Second, if $\int_{\mathbb{R}^n} (1+|\vec{x}|_2^2)^l |f|^2 \, d\vec{x} < \infty$, then its Fourier transform $\hat{f}$ possesses weak partial derivatives $D_{\vec{k}}^{\vec{\alpha}} \hat{f}$ for all $|\vec{\alpha}|_1 \leq l$ and the relations $D_{\vec{k}}^{\vec{\alpha}} \hat{f} = (-1)^{|\vec{\alpha}|_1} \mathcal{F}[\vec{x}^{\vec{\alpha}} f]$ are given with the usual distributive meaning. Finally, if we suppose that $f \in \mathcal{L}^2$ shows $\mathcal{L}^2$-decay of order $l + \frac{n+\epsilon}{2}$, i.e. $\int_{\mathbb{R}^n} |f(\vec{x})|^2 (1+\vec{x})^{2l+n+\epsilon} \, d\vec{x} < \infty$ for $\epsilon > 0$, $l \in \mathbb{N}$, it follows that

$$\int_{\mathbb{R}^n} |f(\vec{x})| (1+|\vec{x}|_2)^l \, d\vec{x} \leq \sqrt{\int_{\mathbb{R}^n} |f(\vec{x})|^2 (1+|\vec{x}|_2)^{2l+n+\epsilon} \, d\vec{x}} \sqrt{\int_{\mathbb{R}^n} (1+|\vec{x}|_2)^{-(n+\epsilon)} \, d\vec{x}} < \infty$$

and thus $\hat{f}$ possesses l continuous derivatives.

2.2 Many-particle Sobolev spaces

Based on the relations discussed in the previous Section 2.1.3 it is common to characterize the smoothness classes of a function f by the decay properties of its Fourier transform $\hat{f}$. In that way let us now define isotropic Sobolev spaces in D dimensions.

Definition 2.1. *For $r \in \mathbb{R}$, $D \in \mathbb{N}$ we define*

$$\mathcal{H}^r(\mathbb{R}^D) := \left\{ f \in \mathcal{S}'(\mathbb{R}^D) \, : \, w^r \hat{f} \in \mathcal{L}^2(\mathbb{R}^D) \right\},$$

with the positive weight

$$w : \mathbb{R}^D \to \mathbb{R} : \mathbf{k} \mapsto \sqrt{1+|\mathbf{k}|_2^2}. \tag{2.6}$$

---
[1] Let us remark that the space $\mathcal{D}(\mathbb{R}^n)$ of all infinitely differentiable functions with bounded support is dense in the Schwartz space $\mathcal{S}(\mathbb{R}^n)$. In particular $\mathcal{S}(\mathbb{R}^n)$ is larger than $\mathcal{D}(\mathbb{R}^n)$, e.g. the Gaussian $e^{-|\vec{x}|_2^2}$ is not in $\mathcal{D}(\mathbb{R}^n)$ but it is in $\mathcal{S}(\mathbb{R}^n)$, whereas we have for its duals $\mathcal{S}' \subset \mathcal{D}'$.

2.2 Many-particle Sobolev spaces

In particular, the Sobolev space $\mathcal{H}^r(\mathbb{R}^D)$ is a Hilbert space together with the Hermitian inner product

$$\langle f, g \rangle_{\mathcal{H}^r} := \langle w^r \hat{f}, w^r \hat{g} \rangle_{\mathcal{L}^2} = \int_{\mathbb{R}^D} (1 + |\mathbf{k}|_2^2)^r \hat{f}^*(\mathbf{k}) \hat{g}(\mathbf{k}) \, d\mathbf{k}$$

and the norm

$$\|f\|_{\mathcal{H}^r} := \sqrt{\langle f, f \rangle_{\mathcal{H}^r}} = \sqrt{\int_{\mathbb{R}^D} (1 + |\mathbf{k}|_2^2)^r |\hat{f}(\mathbf{k})|^2 \, d\mathbf{k}},$$

where $\mathcal{S} \subset \mathcal{H}^r \subset \mathcal{S}'$ and $\mathcal{S}$ is dense in $\mathcal{L}^2$. Let us remark that for $r \in \mathbb{N}_0$ the space $\mathcal{H}^r$ possesses the equivalent norm

$$\|f\|_{\mathcal{H}^r} := \sqrt{\sum_{|\vec{\alpha}|_1 \leq r} \langle D^{\vec{\alpha}} f, D^{\vec{\alpha}} f \rangle}. \tag{2.7}$$

Consequently Definition 2.1 naturally extends the classical Sobolev spaces[2] for $r \in \mathbb{N}_0$ to $r \in \mathbb{R}$. Moreover, we introduce the $\mathcal{H}^r$-seminorm via

$$|f|_{\mathcal{H}^r} := \sqrt{\int_{\mathbb{R}^D} |\mathbf{k}|_2^{2r} |\hat{f}(\mathbf{k})|^2 \, d\mathbf{k}}. \tag{2.8}$$

Now, starting from the one-particle space $\mathcal{H}^r(\mathbb{R}^D)$ we build Sobolev spaces for N particles. Obviously there are many ways to generalize the concept of Sobolev spaces [1] from the one-particle case to the higher dimensional many-particle case. Two simple possibilities are the additive or multiplicative combination, i.e. an arithmetic or geometric averaging of the frequencies for the different particles. We use the following definition that combines both approaches. We denote the weights

$$w_{\text{iso}} : (\mathbb{R}^D)^N \to \mathbb{R} : \vec{\mathbf{k}} \mapsto \sqrt{1 + \sum_{p=1}^N (\omega(\mathbf{k}_p))^2} \tag{2.9}$$

and

$$w_{\text{mix}} : (\mathbb{R}^D)^N \to \mathbb{R} : \vec{\mathbf{k}} \mapsto \sqrt{\prod_{p=1}^N (1 + (\omega(\mathbf{k}_p))^2)}, \tag{2.10}$$

where we set ω to the Euclidean norm in $\mathbb{R}^D$, i.e.

$$\omega : \mathbb{R}^D \to \mathbb{R} : \mathbf{k} \mapsto |\mathbf{k}|_2. \tag{2.11}$$

Definition 2.2. *For $-\infty < t, r < \infty$, $D \in \mathbb{N}$, $N \in \mathbb{N}$ we set*

$$\mathcal{H}_{\text{mix}}^{t,r}((\mathbb{R}^D)^N) := \left\{ f \in \mathcal{S}'((\mathbb{R}^D)^N) \,:\, w_{\text{mix}}^t w_{\text{iso}}^r \hat{f} \in \mathcal{L}^2((\mathbb{R}^D)^N) \right\}$$

with the norm

$$\|f\|_{\mathcal{H}_{\text{mix}}^{t,r}} := \sqrt{\int_{(\mathbb{R}^D)^N} (w_{\text{mix}}(\vec{\mathbf{k}}))^{2t} (w_{\text{iso}}(\vec{\mathbf{k}}))^{2r} |\hat{f}(\vec{\mathbf{k}})|^2 \, d\vec{\mathbf{k}}}.$$

---

[2]Note that since $\mathcal{S}$ is dense in $\mathcal{L}^2$, the classical Sobolev spaces $\mathcal{H}^r$, $r \in \mathbb{N}_0$, may also be considered as the completion of $\mathcal{S}$ under the corresponding norm (2.7).

2 Sobolev Spaces for Many-Particle Functions

In particular, the Sobolev space $\mathcal{H}_{\text{mix}}^{t,r}((\mathbb{R}^D)^N)$ is a Hilbert space together with the Hermitian inner product

$$\langle f, g \rangle_{\mathcal{H}_{\text{mix}}^{t,r}} := \langle w_{\text{mix}}^t w_{\text{iso}}^r \hat{f}, w_{\text{mix}}^t w_{\text{iso}}^r \hat{g} \rangle_{\mathcal{L}^2}$$

$$= \int_{(\mathbb{R}^D)^N} (w_{\text{mix}}(\vec{\mathbf{k}}))^{2t} (w_{\text{iso}}(\vec{\mathbf{k}}))^{2r} \hat{f}^*(\mathbf{k}) \hat{g}(\mathbf{k}) \, d\mathbf{k}.$$

Note that due to equation (2.5), the squared weight w_{mix}^2 relates to the operator $\prod_{p=1}^{N}(1 - \Delta_p)$. Here, the operator Δ_p is the Laplacian acting on the $\mathbf{x}_p$-component, i.e. $\Delta_p = \sum_{d=1}^{D} \partial^2 / \partial (x_{p,(d)})^2$. It expresses the multiplicative combination of the $\mathcal{H}^1(\mathbb{R}^D)$-norm of the one-particle space with a norm of the N-particle space which involves mixed derivatives. Furthermore, w_{iso}^2 relates to the operator $1 - \sum_{p=1}^{N} \Delta_p$ and directly creates an associated $\mathcal{H}^1((\mathbb{R}^D)^N)$-norm for the N-particle space. A t- and r-times application of these operators leads to the corresponding multiplicative combination of the $\mathcal{H}^t(\mathbb{R}^D)$-norm and the $\mathcal{H}^r((\mathbb{R}^D)^N)$-norm, respectively. Moreover, for $r, l \in \mathbb{N}_0$ the space $\mathcal{H}_{\text{mix}}^{t,r}$ possesses according to (2.5) the equivalent norm given by

$$\|f\|_{\mathcal{H}_{\text{mix}}^{t,r}}^2 := \sum_{\max_{p=1}^{N} |\mathbf{a}_p|_1 \leq t} \sum_{\sum_{p=1}^{N} |\mathbf{b}_p|_1 \leq r} \|D^{\vec{\mathbf{a}}+\vec{\mathbf{b}}} f\|_{\mathcal{L}^2}^2. \qquad (2.12)$$

In particular, the standard isotropic Sobolev spaces [1] as well as the Sobolev spaces of dominating mixed smoothness [160, 182], both generalized to the N-particle case [67, 69], are included in the definition of $\mathcal{H}_{\text{mix}}^{t,r}$. They can be written as

$$\mathcal{H}^r((\mathbb{R}^D)^N) = \mathcal{H}_{\text{mix}}^{0,r}((\mathbb{R}^D)^N) \quad \text{and} \quad \mathcal{H}_{\text{mix}}^t((\mathbb{R}^D)^N) = \mathcal{H}_{\text{mix}}^{t,0}((\mathbb{R}^D)^N),$$

respectively. Hence, the parameter r from Definition 2.2 governs the isotropic smoothness, whereas t governs the mixed smoothness. Thus, the spaces $\mathcal{H}_{\text{mix}}^{t,r}$ give us a quite flexible framework for the study of problems in Sobolev spaces.

Note that if the conditions $t' + r' \leq t + r$ and $Nt' + r' \leq Nt + r$ are fulfilled, then the embedding

$$\mathcal{H}_{\text{mix}}^{t,r}((\mathbb{R}^D)^N) \subset \mathcal{H}_{\text{mix}}^{t',r'}((\mathbb{R}^D)^N) \qquad (2.13)$$

holds. Thus, relation (2.13) holds either for $t' + r' \leq t + r$, $t - t' \geq 0$ or for $Nt' + r' \leq Nt + r$, $t - t' < 0$. Especially, the inclusions $\mathcal{H}_{\text{mix}}^t \subset \mathcal{H}^t \subset \mathcal{H}_{\text{mix}}^{t/N}$ for $t \geq 0$ and $\mathcal{H}_{\text{mix}}^{t/N} \subset \mathcal{H}^t \subset \mathcal{H}_{\text{mix}}^t$ for $t \leq 0$ hold.

Let us remark that the spaces $\mathcal{H}_{\text{mix}}^{t,r}((\mathbb{R}^D)^N)$ may be written as a mixture of tensor products of one-particle Sobolev spaces: Let $t \in \mathbb{R}$, $r \in \mathbb{R}_0^+$, $\vec{\mathbf{1}} = (1, \ldots, 1) \in \mathbb{R}^N$ and $\vec{e}_i = (0, \ldots, 0, 1, 0, \ldots, 0)$ the i-th unit-vector in $\mathbb{R}^N$. Then we have

$$\mathcal{H}_{\text{mix}}^{t,r}((\mathbb{R}^D)^N) = \mathcal{H}_{\text{mix}}^{t\vec{\mathbf{1}}+r\vec{e}_1}((\mathbb{R}^D)^N) \cap \cdots \cap \mathcal{H}_{\text{mix}}^{t\vec{\mathbf{1}}+r\vec{e}_N}((\mathbb{R}^D)^N),$$

where

$$\mathcal{H}_{\text{mix}}^{\vec{k}}((\mathbb{R}^D)^N) := \mathcal{H}^{k_1}(\mathbb{R}^D) \otimes \cdots \otimes \mathcal{H}^{k_N}(\mathbb{R}^D).$$

This can be easily deduced from the definition of the tensor product by orthonormal systems [186] and the intersection of spaces; compare [67, 72, 73, 75, 93]. In particular, it follows that $\mathcal{H}_{\text{mix}}^t((\mathbb{R}^D)^N)$ and $\mathcal{H}^r((\mathbb{R}^D)^N)$ can be written as

$$\mathcal{H}_{\text{mix}}^{t,0}((\mathbb{R}^D)^N) = \mathcal{H}^t(\mathbb{R}^D) \otimes \cdots \otimes \mathcal{H}^t(\mathbb{R}^D) = \mathcal{H}_{\text{mix}}^t((\mathbb{R}^D)^N) \qquad (2.14)$$

2.2 Many-particle Sobolev spaces

and
$$\mathcal{H}_{\text{mix}}^{0,r}((\mathbb{R}^D)^N) = \mathcal{H}^{r\vec{e}_1}(\mathbb{R}^D) \cap \cdots \cap \mathcal{H}^{r\vec{e}_N}(\mathbb{R}^D) = \mathcal{H}^r((\mathbb{R}^D)^N),$$

respectively. In the case of $r < 0$ a space $\mathcal{H}_{\text{mix}}^{t,r}((\mathbb{R}^D)^N)$ can be defined as the dual space of $\mathcal{H}_{\text{mix}}^{-t,-r}((\mathbb{R}^D)^N)$, i.e. $\mathcal{H}_{\text{mix}}^{t,r}((\mathbb{R}^D)^N) = \left(\mathcal{H}_{\text{mix}}^{-t,-r}((\mathbb{R}^D)^N)\right)'$. Note that similar results hold for problems on $([0,1]^D)^N$ with Dirichlet, Neumann or periodic boundary conditions and certain cases of mixed boundary conditions [72, 73, 75, 93, 107].

Let us remark that the simple norm equivalence

$$\sum_{p=1}^N \omega(\mathbf{k}_p) \simeq \max_{p=1,\ldots,N} \omega(\mathbf{k}_p)$$

holds, where the constant in the upper estimate involves a factor of N. This allows us to switch from w_{iso} as in (2.9) to

$$w_{\text{iso}}(\vec{\mathbf{k}}) := \sqrt{1 + \max_{p=1,\ldots,N} (\omega(\mathbf{k}_p))^2}.$$

With basically the same norm equivalence we can replace (2.11) by

$$\omega(\mathbf{k}) = |\mathbf{k}|_\infty. \qquad (2.15)$$

These changes in the definitions of ω and w_{iso} result in the same spaces $\mathcal{H}_{\text{mix}}^{t,r}((\mathbb{R}^D)^N)$. In the following we work with these equivalent definitions, i.e.

$$w_{\text{iso}} : (\mathbb{R}^D)^N \to \mathbb{R} : \vec{\mathbf{k}} \mapsto \sqrt{1 + \max_{p=1,\ldots,N} |\mathbf{k}_p|_\infty^2} \qquad (2.16)$$

and

$$w_{\text{mix}} : (\mathbb{R}^D)^N \to \mathbb{R} : \vec{\mathbf{k}} \mapsto \sqrt{\prod_{p=1}^N (1 + |\mathbf{k}_p|_\infty^2)}, \qquad (2.17)$$

since they simplify error estimates and complexity substantially.

Furthermore, let us shortly discuss the case of periodic Sobolev spaces. Let $\mathbb{T}^n$ be the n-torus, which is identified with the n-dimensional cube $\mathbb{T}^n \subset \mathbb{R}^n$, $\mathbb{T} = [-\pi, \pi]$, where opposite points identified. For $-\infty < t, r < \infty$, $D \in \mathbb{N}$, $N \in \mathbb{N}$ we introduce

$$\mathcal{H}_{\text{mix}}^{t,r}((\mathbb{T}^D)^N) := \left\{ f = \sum_{\vec{\mathbf{k}} \in (\mathbb{Z}^D)^N} c_{\vec{\mathbf{k}}} \tilde{e}_{\vec{\mathbf{k}}} \in \mathcal{D}'((\mathbb{T}^D)^N) : \right.$$
$$\left. \left\{ w_{\text{mix}}(\vec{\mathbf{k}}))^t (w_{\text{iso}}(\vec{\mathbf{k}}))^r c_{\vec{\mathbf{k}}} \right\}_{\vec{\mathbf{k}} \in (\mathbb{Z}^D)^N} \in \ell^2 \right\}$$

analogously to Definition 2.2 and the norm

$$\|f\|_{\mathcal{H}_{\text{mix}}^{t,r}((\mathbb{T}^D)^N)} := \sqrt{\sum_{\vec{\mathbf{k}} \in (\mathbb{Z}^D)^N} (w_{\text{mix}}(\vec{\mathbf{k}}))^{2t} (w_{\text{iso}}(\vec{\mathbf{k}}))^{2r} |c_{\vec{\mathbf{k}}}|^2},$$

where the counterpart of the Fourier transform are the Fourier coefficients given for $f \in \mathcal{D}'((\mathbb{T}^D)^N)$, $\vec{\mathbf{k}} \in (\mathbb{Z}^D)^N$ by

$$c_{\vec{\mathbf{k}}} = \int_{(\mathbb{T}^D)^N} \tilde{e}_{\vec{\mathbf{k}}}^*(\vec{\mathbf{x}}) f(\vec{\mathbf{x}}) \, d\vec{\mathbf{x}} = \langle \tilde{e}_{\vec{\mathbf{k}}}, f \rangle_{\mathcal{L}^2((\mathbb{T}^D)^N)}.$$

2 Sobolev Spaces for Many-Particle Functions

Here, $\tilde{e}_{\vec{k}}$ denotes the orthonormal Fourier basis written in the form

$$\tilde{e}_{\vec{k}} := \bigotimes_{p=1}^{N} \tilde{e}_{\mathbf{k}_p}, \quad \tilde{e}_{\mathbf{k}}(x) := \bigotimes_{d=1}^{D} \tilde{e}_{k_d}, \quad \tilde{e}_k(x) := \frac{1}{\sqrt{2\pi}} e^{ikx}, \qquad (2.18)$$

which especially is a complete orthonormal system of $\mathcal{L}^2((\mathbb{T}^D)^N)$. The role of the Schwartz space $\mathcal{S}((\mathbb{R}^D)^N)$ is taken over by the space $\mathcal{D}((\mathbb{T}^D)^N)$ of C^∞ functions on $(\mathbb{T}^D)^N$, and the role of the space $\mathcal{S}'((\mathbb{R}^D)^N)$ of tempered distributions on $(\mathbb{R}^D)^N$ is taken over by the space $\mathcal{D}'((\mathbb{T}^D)^N)$ of periodic distributions [160].

2.3 Hyperbolic cross approximation

In the following we shortly discuss the approximation of a function in $\mathcal{H}_{\text{mix}}^{t,r}((\mathbb{R}^D)^N)$ by a function which is band-limited to a general hyperbolic cross domain.

Definition 2.3. *For $T < 1$ and $K \geq 1$ we define the general hyperbolic cross by*

$$\mathcal{K}_K^T := \left\{ \vec{\mathbf{k}} \in (\mathbb{R}^D)^N : w_{\text{mix}}(\vec{\mathbf{k}})(w_{\text{iso}}(\vec{\mathbf{k}}))^{-T} \leq K^{1-T} \right\},$$

with the natural extension to the case of $T \to -\infty$ by

$$\mathcal{K}_K^{-\infty} := \left\{ \vec{\mathbf{k}} \in (\mathbb{R}^D)^N : w_{\text{iso}}(\vec{\mathbf{k}}) \leq K \right\}.$$

We further define the general hyperbolic cross space $\mathcal{V}_{\mathcal{K}_K^T}$ by the space of functions with vanishing Fourier transforms outside the domain $\mathcal{K}_K^T$.

The parameter T allows us to switch from the isotropic full space case $T = -\infty$ to the conventional anisotropic hyperbolic cross space case $T = 0$, compare [24, 72, 73, 106], and with $T \in (0,1)$ also allows to create subspaces of the conventional hyperbolic cross space. It can be easily deduced from Definition 2.3 that the inclusions

$$\mathcal{K}_K^{T_1} \subset \mathcal{K}_K^{T_2} \subset \mathcal{K}_K^{-\infty}, \qquad (2.19)$$

$\mathcal{K}_K^{T_2} \subset \mathcal{K}_{\tilde{K}}^{T_1}$ with $\tilde{K} = K^{\frac{(1-T_2)(N-T_1)}{(N-T_2)(1-T_1)}}$ and $\mathcal{K}_K^{-\infty} \subset \mathcal{K}_{\tilde{\tilde{K}}}^{T_1}$ with $\tilde{\tilde{K}} = K^{\frac{N-T_1}{1-T_1}}$ (2.20)

for $K \geq 1$, $-\infty < T_2 \leq T_1 < 1$ hold. Also, for $1 \leq K_1 \leq K_2$, $-\infty < T < 1$ we have the inclusion $\mathcal{K}_{K_1}^T \subset \mathcal{K}_{K_2}^T$. Figure 2.1(a) displays sets $\mathcal{K}_K^T$ for various choices of T for the case of $D = 1$, $N = 2$ and $K = 10$, and Figure 2.1(b) displays some sets according to inclusion (2.20).

The projection of a function f onto the general hyperbolic cross space $\mathcal{V}_{\mathcal{K}_K^T}$ is given with the help of the characteristic function $\chi_{\mathcal{K}_K^T}$ of the domain $\mathcal{K}_K^T$ by

$$\begin{aligned}\mathcal{P}_{\mathcal{K}_K^T}[f](\vec{\mathbf{x}}) &= \mathcal{F}^{-1}[\chi_{\mathcal{K}_K^T}\hat{f}](\vec{\mathbf{x}}),\\ &= \left(\frac{1}{\sqrt{2\pi}}\right)^{DN} \int_{\mathcal{K}_K^T} e^{i\vec{\mathbf{x}}^T\vec{\mathbf{k}}} \hat{f}(\vec{\mathbf{k}})\, d\vec{\mathbf{k}},\end{aligned} \qquad (2.21)$$

where we also write $f_{\mathcal{K}_K^T} := \mathcal{P}_{\mathcal{K}_K^T}[f]$.

2.3 Hyperbolic cross approximation

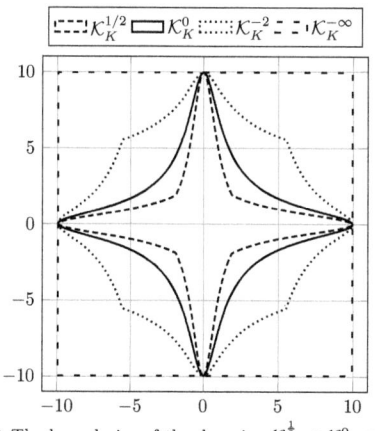
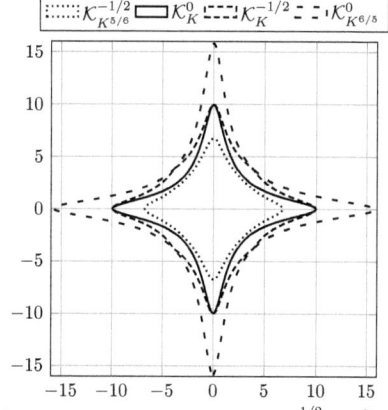

(a) The boundaries of the domains $\mathcal{K}_K^{\frac{1}{2}} \subset \mathcal{K}_K^0 \subset \mathcal{K}_K^{-2} \subset \mathcal{K}_K^{-\infty}$ for $K = 10$.

(b) The boundaries of the domains $\mathcal{K}_{K^{5/6}}^{-1/2} \subset \mathcal{K}_K^0 \subset \mathcal{K}_K^{-1/2} \subset \mathcal{K}_{K^{6/5}}^0$ for $K = 10$.

Figure 2.1: General hyperbolic cross sets $\mathcal{K}_K^T$ according to Definition 2.3 are depicted for $D = 1$ and $N = 2$. Here, w_{iso} and w_{mix} are given by (2.16) and (2.17), respectively.

Lemma 2.1. *Let* $t' + r' < t + r$, $t - t' \geq 0$, $f \in \mathcal{H}_{\text{mix}}^{t,r}((\mathbb{R}^D)^N)$ *and* $f_{\mathcal{K}_K^T}$ *the projection of f onto* $\mathcal{V}_{\mathcal{K}_K^T}$. *Then for* $t - t' > 0$ *it holds*

$$\|f - f_{\mathcal{K}_K^T}\|_{\mathcal{H}_{\text{mix}}^{t',r'}} \leq \begin{cases} K^{(r'-r)-(t-t')+(T(t-t')-(r'-r))\frac{N-1}{N-T}} \|f\|_{\mathcal{H}_{\text{mix}}^{t,r}} & \text{for } T \geq \frac{r'-r}{t-t'}, \\ K^{(r'-r)-(t-t')} \|f\|_{\mathcal{H}_{\text{mix}}^{t,r}} & \text{for } T \leq \frac{r'-r}{t-t'}, \end{cases} \quad (2.22)$$

and for $t - t' = 0$ *it holds*

$$\|f - f_{\mathcal{K}_K^T}\|_{\mathcal{H}_{\text{mix}}^{t',r'}} \leq \begin{cases} K^{(r'-r)\left(1-\frac{N-1}{N-T}\right)} \|f\|_{\mathcal{H}_{\text{mix}}^{t,r}} & \text{for } T > -\infty, \\ K^{(r'-r)} \|f\|_{\mathcal{H}_{\text{mix}}^{t,r}} & \text{for } T = -\infty. \end{cases} \quad (2.23)$$

Proof. We denote the complement set of $\mathcal{K}_K^T$ in $(\mathbb{R}^D)^N$ by $\overline{\mathcal{K}_K^T} := (\mathbb{R}^D)^N \setminus \mathcal{K}_K^T$. Due to (2.21) the Fourier transform of $f_{\mathcal{K}_K^T}$ is given by $\chi_{\mathcal{K}_K^T} \hat{f}$. Thus, we directly obtain

$$\begin{aligned}
\|f - f_{\mathcal{K}_K^T}\|_{\mathcal{H}_{\text{mix}}^{t',r'}}^2 &= \int_{\overline{\mathcal{K}_K^T}} (w_{\text{mix}}(\vec{k}))^{2t'} (w_{\text{iso}}(\vec{k}))^{2r'} |\hat{f}(\vec{k})|_2^2 \, d\vec{k} \\
&= \int_{\overline{\mathcal{K}_K^T}} \frac{(w_{\text{iso}}(\vec{k}))^{2(r'-r)}}{(w_{\text{mix}}(\vec{k}))^{2(t-t')}} (w_{\text{mix}}(\vec{k}))^{2t} (w_{\text{iso}}(\vec{k}))^{2r} |\hat{f}(\vec{k})|^2 \, d\vec{k} \\
&\leq \left(\sup_{\vec{k} \in \overline{\mathcal{K}_K^T}} \frac{(w_{\text{iso}}(\vec{k}))^{2(r'-r)}}{(w_{\text{mix}}(\vec{k}))^{2(t-t')}} \right) \int_{\overline{\mathcal{K}_K^T}} (w_{\text{mix}}(\vec{k}))^{2t} (w_{\text{iso}}(\vec{k}))^{2r} |\hat{f}(\vec{k})|^2 \, d\vec{k} \\
&\leq \left(\sup_{\vec{k} \in \overline{\mathcal{K}_K^T}} \frac{(w_{\text{iso}}(\vec{k}))^{2(r'-r)}}{(w_{\text{mix}}(\vec{k}))^{2(t-t')}} \right) \|f\|_{\mathcal{H}_{\text{mix}}^{t,r}}. \quad (2.24)
\end{aligned}$$

17

Now, evaluating the supremum in (2.24) with the help of Definition 2.3 of the set $\mathcal{K}_K^T$ and with the help of either inclusion (2.19) with $T_1 = \frac{r'-r}{t-t'}$, $T_2 = T$ for the case of $T \leq \frac{r'-r}{t-t'}$ or inclusion (2.20) with $T_1 = T$, $T_2 = \frac{r'-r}{t-t'}$ for the case of $T \geq \frac{r'-r}{t-t'}$ leads to the desired result for the case of $t - t' > 0$ in (2.22). In the same way, estimate (2.23) for the case of $t - t' = 0$ is obtained with the help of inclusion (2.19) for $T = -\infty$ and inclusion (2.20) with $T_1 = T$ for $T < -\infty$. □

This type of estimate was already given for the case of measuring the error in the $\mathcal{H}^{0,r'}$-norm (i.e. $t' = 0$) and a dyadically refined wavelet basis with $D = 1$ for the periodic case on a finite domain in [72, 73, 106, 107]. It is a generalization of the energy-norm based sparse grid approach of [23, 24, 64] where the case of $t' = 0$, $r' = 1$, $t = 2$, $r = 0$ was considered using a hierarchical piecewise linear basis.

Let us discuss some cases for measuring the error in the $\mathcal{H}^{0,r'}$-norm (i.e. $t' = 0$). For the standard Sobolev space $\mathcal{H}_\mathrm{mix}^{0,r}$ (i.e. $t = 0$, $0 \leq r' < r$) and the spaces $\mathcal{V}_{\mathcal{K}_K^T}$ with $T > -\infty$ the resulting order is dependent on T and dependent on the number of particles N. In particular, the order even deteriorates with larger T. For the standard Sobolev spaces of bounded mixed derivatives $\mathcal{H}_\mathrm{mix}^{t,0}$ (i.e. $r = 0$, $0 \leq r' < t$) and the spaces $\mathcal{V}_{\mathcal{K}_K^T}$ with $T > \frac{r'}{t}$ the resulting order is dependent on T and on the number of particles N whereas for $T \leq \frac{r'}{t}$ the resulting order is independent of T and N. If for example we restrict the class of functions to $\mathcal{H}_\mathrm{mix}^{t,r}$ (i.e. $t > 0$, $r \geq 0$) and measure the error in the $\mathcal{H}_\mathrm{mix}^{0,r}$-norm (i.e. $r' = r$), the approximation order is dependent on N for all $T > 0$ and independent on N and T for all $T \leq 0$.

Let us finally note that Lemma 2.1 can easily be extended to the case of employing w_iso, w_mix and ω given by (2.9), (2.10) and (2.11) in Definition 2.2 and 2.3. Figure 2.2 displays some example sets.

2.3.1 Multiscale decomposition of general hyperbolic cross spaces

To obtain a multiscale decomposition of general hyperbolic cross spaces we use a Littlewood-Paley like dyadic partition of unity to decompose the Fourier space [57, 139]. To this end we consider the following index sets which are introduced in [72]:

Definition 2.4. *For $T < 1$ and $L \in \mathbb{N}$ we define the parametric index sets*

$$\mathcal{I}_L^T := \left\{ \vec{l} \in \mathbb{N}^N \ : \ |\vec{l}|_1 - T|\vec{l}|_\infty \leq L(1-T) + (N-1) \right\},$$

with the natural extension to the case of $T \to -\infty$ by

$$\mathcal{I}_L^{-\infty} := \left\{ \vec{l} \in \mathbb{N}^N \ : \ |\vec{l}|_\infty \leq L \right\}.$$

With regard to the cardinality of a set $\mathcal{I}_L^T$, the following lemma is shown in [72, 73]:

Lemma 2.2. *It holds*

$$\sum_{\vec{l} \in \mathcal{I}_L^T} 2^{|\vec{l}|_1 - N} \leq \begin{cases} \frac{N}{2}\left(1 - 2^{-\frac{1}{1/T-1}}\right)^{-N} 2^L = \mathcal{O}(2^L) & \text{for } 0 < T < 1, \\ \mathcal{O}(2^{L\frac{T-1}{T/N-1}}) & \text{for } -\infty < T < 0, \\ \mathcal{O}(2^{LN}) & \text{for } T = -\infty. \end{cases}$$

2.3 Hyperbolic cross approximation

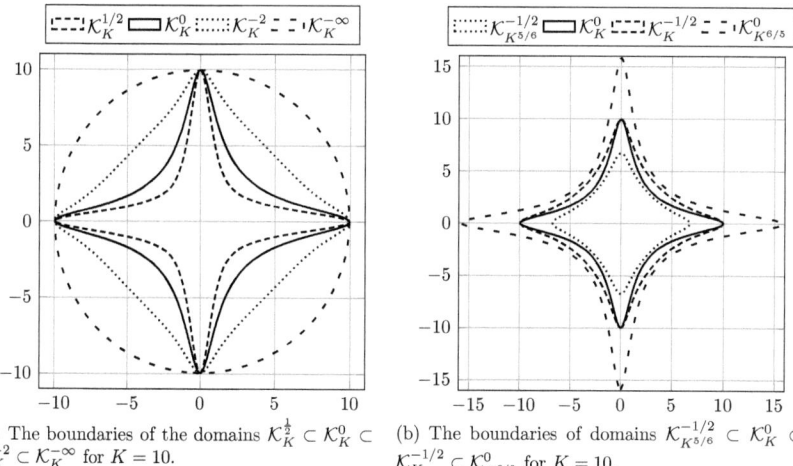

Figure 2.2: General hyperbolic cross sets $\mathcal{K}_K^T$ according to Definition 2.3 are depicted for $D = 1$ and $N = 2$. Here, w_{iso} and w_{mix} are given by (2.9) and (2.10), respectively.

The case of $T = 0$ is covered by the additional estimate

$$\sum_{\vec{l} \in \mathcal{I}_L^T} 2^{|\vec{l}|_1 - N} \leq 2^L \left(\frac{L^{N-1}}{(N-1)!} + \mathcal{O}(L^{N-2}) \right) = \mathcal{O}(2^L L^{N-1}),$$

for $0 \leq T \leq 1/L$.

Now, for $\mathcal{I}_L^T$ according to Definition 2.4, we introduce a general hyperbolic cross

$$\mathcal{K}_{\mathcal{I}_L^T} := \bigcup_{\vec{l} \in \mathcal{I}_L^T} Q_{\vec{l}}, \quad Q_{\vec{l}} := Q_{l_1} \times \cdots \times Q_{l_N}, \tag{2.25}$$

where we define the domain $Q_l \subset \mathbb{R}^D$ by

$$Q_l := \left\{ \mathbf{k} \in \mathbb{R}^D : 1 + (\omega(\mathbf{k}))^2 \leq 2^{2l} \right\} \tag{2.26}$$

for an index $l \in \mathbb{N}$. Note that for $l \in \mathbb{N}$ the domain

$$\tilde{Q}_l := \left\{ \mathbf{k} \in \mathbb{R}^D : |\mathbf{k}|_\infty \leq 2^l \right\} \tag{2.27}$$

covers the Q_l domain and thus it holds the inclusion

$$\mathcal{K}_{\mathcal{I}_L^T} \subset \tilde{\mathcal{K}}_{\mathcal{I}_L^T} := \bigcup_{\vec{l} \in \mathcal{I}_L^T} \tilde{Q}_{\vec{l}},$$

with $\tilde{Q}_{\vec{l}} := \tilde{Q}_{l_1} \times \cdots \times \tilde{Q}_{l_N}$. Moreover, we have the following relation:

2 Sobolev Spaces for Many-Particle Functions

Lemma 2.3. *Let $T < 1$, $K \geq 1$, $\mathcal{K}_K^T$ defined according to Definition 2.3 and $\mathcal{K}_{\mathcal{I}_L^T}$ as in (2.25) with $L = \lceil \log_2(K) \rceil$, then it holds the inclusion*

$$\mathcal{K}_K^T \subset \mathcal{K}_{\mathcal{I}_L^T}.$$

Proof. Let $\vec{k} \in \mathcal{K}_K^T$. Then, by definition

$$\left(\prod_{p=1}^N \left(1 + (\omega(\mathbf{k}_p))^2\right) \right) \left(1 + \max_{1 \leq p \leq N} (\omega(\mathbf{k}_p))^2 \right)^{-T} \leq K^{2(1-T)}.$$

Now, for $p = 1, \ldots, N$ we set

$$l_p := \left\lfloor 1 + \frac{1}{2} \log_2 \left(1 + (\omega(\mathbf{k}_p))^2\right) \right\rfloor \quad \text{and} \quad \Delta l_p := 1 + \frac{1}{2} \log_2 \left(1 + (\omega(\mathbf{k}_p))^2\right) - l_p.$$

Then we have $1 + (\omega(\mathbf{k}_p))^2 = 2^{2(l_p - 1 + \Delta l_p)}$, $l_p \in \mathbb{N}$, $\Delta l_p \in [0,1)$, and there follows the relation

$$\left(\prod_{p=1}^N 2^{2(l_p - 1 + \Delta l_p)} \right) \left(\max_{1 \leq p \leq N} 2^{2(l_p - 1 + \Delta l_p)} \right)^{-T} \leq 2^{2 \log_2(K)(1-T)}.$$

With $p_m = \arg\max_{p=1,\ldots,N} (\omega(\mathbf{k}_p))^2$ we obtain

$$|\vec{l}|_1 - T|\vec{l}|_\infty \leq \log_2(K)(1-T) + (N-1) - \left(\sum_{p \neq p_m}^N \Delta l_p + \Delta l_{p_m}(1-T) \right)$$

$$\leq \log_2(K)(1-T) + (N-T)$$
$$\leq \log_2(K)(1-T) + (N-1)$$
$$\leq \lceil \log_2(K) \rceil (1-T) + (N-1)$$
$$= L(1-T) + (N-1),$$

which completes the proof. □

In order to obtain a multiscale decomposition of the general hyperbolic cross spaces $\mathcal{V}_{\mathcal{K}_K^T}$, we use a smooth dyadic decomposition of unity. To this end, for $l \in \mathbb{N}$ and Q_l as in (2.26) let $\eta_{Q_l} : \mathbb{R}^D \to [0,1]$ an infinitely differentiable function such that $\eta_{Q_l}(\mathbf{k}) = 1$ if $\mathbf{k} \in Q_l$, and $\operatorname{supp} \eta_{Q_l} \subset Q_{l+1}$, i.e. a so-called bump function. Then we put

$$\eta_1 := \eta_{Q_1} \quad \text{and} \quad \eta_l := \eta_{Q_l} - \eta_{Q_{l-1}}, l \in \{2, 3, 4, \ldots\} \tag{2.28}$$

and obtain with the help of the tensor product functions

$$\eta_{\vec{l}} := \bigotimes_{p=1}^N \eta_{l_p}, \quad \vec{l} \in \mathbb{N}^N, \tag{2.29}$$

a partition of unity

$$\sum_{\vec{l} \in \mathbb{N}^N} \eta_{\vec{l}}(\vec{\mathbf{k}}) = 1, \quad \vec{\mathbf{k}} \in (\mathbb{R}^D)^N. \tag{2.30}$$

Hence, every $f \in \mathcal{S}'$ splits into parts

$$f_{\vec{l}}(\vec{x}) := \mathcal{F}^{-1}[\eta_{\vec{l}} \mathcal{F} f](\vec{x}), \tag{2.31}$$

2.3 Hyperbolic cross approximation

i.e. it holds the relation
$$f(\vec{x}) = \sum_{\vec{l} \in \mathbb{N}^N} f_{\vec{l}}(\vec{x}), \tag{2.32}$$

for $f \in \mathcal{S}'$ (with the usual distributive meaning), similar to a variant of Calderón representation formula in the case of inhomogeneous spaces [45, 57, 139]. Let us now consider the partial sum

$$f_{\mathcal{I}_L^T}(\vec{x}) := \sum_{\vec{l} \in \mathcal{I}_L^T} f_{\vec{l}}(\vec{x}), \tag{2.33}$$

which corresponds to a hyperbolic cross index set $\mathcal{I}_L^T$ of a multiscale decomposition (2.32) of a function $f \in \mathcal{H}_{\text{mix}}^{t,r}$. Here, the domain $\mathcal{K}_{\mathcal{I}_L^T}$ covers the general hyperbolic cross $\mathcal{K}_{2^L}^T$ due to Lemma 2.3 and by definition we have $\sum_{\vec{l} \in \mathcal{I}_L^T} \eta_{\vec{l}}(\vec{k}) = 1$ for $\vec{k} \in \mathcal{K}_{2^L}^T \subset \mathcal{K}_{\mathcal{I}_L^T}$ and $0 \leq \sum_{\vec{l} \in \mathcal{I}_L^T} \eta_{\vec{l}}(\vec{k}) \leq 1$ for $\vec{k} \in (\mathbb{R}^D)^N \setminus \mathcal{K}_{2^L}^T$. Hence, $f_{\mathcal{I}_L^T}$ approximates f at least as well as the hyperbolic cross projection $f_{\mathcal{K}_{2^L}^T}$ given by (2.21), and analogously to Lemma 2.1 we obtain the following estimates:

Lemma 2.4. Let $t' + r' < t + r$, $t - t' \geq 0$, $f \in \mathcal{H}_{\text{mix}}^{t,r}((\mathbb{R}^D)^N)$ and $f_{\mathcal{I}_L^T} \in \mathcal{V}_{\mathcal{K}_{\mathcal{I}_L^T}}$ as in (2.33). Then for $t - t' > 0$ it holds

$$\|f - f_{\mathcal{I}_L^T}\|_{\mathcal{H}_{\text{mix}}^{t',r'}} \leq \begin{cases} 2^{L\left((r'-r)-(t-t')+(T(t-t')-(r'-r))\frac{N-1}{N-T}\right)} \|f\|_{\mathcal{H}_{\text{mix}}^{t,r}} & \text{for } T \geq \frac{r'-r}{t-t'}, \\ 2^{L((r'-r)-(t-t'))} \|f\|_{\mathcal{H}_{\text{mix}}^{t,r}} & \text{for } T \leq \frac{r'-r}{t-t'}, \end{cases}$$

and for $t - t' = 0$ it holds

$$\|f - f_{\mathcal{I}_L^T}\|_{\mathcal{H}_{\text{mix}}^{t',r'}} \leq \begin{cases} 2^{L(r'-r)\left(1-\frac{N-1}{N-T}\right)} \|f\|_{\mathcal{H}_{\text{mix}}^{t,r}} & \text{for } T > -\infty, \\ 2^{L(r'-r)} \|f\|_{\mathcal{H}_{\text{mix}}^{t,r}} & \text{for } T = -\infty. \end{cases}$$

2.3.2 Approximation by general sparse grid spaces

In the following we shortly discuss an approximation of the parts $f_{\vec{l}}$ as in (2.33), where we assume that the function f is in $\mathcal{H}_{\text{mix}}^{t,r} \subset \mathcal{L}^2$, $r + t \geq 0$, $t \geq 0$ and that its Fourier transform $\hat{f}$ is in $\mathcal{H}_{\text{mix}}^{\hat{t},\hat{r}} \subset \mathcal{L}^2$, $\hat{r} + \hat{t} \geq 0$, $\hat{t} \geq 0$. Let us remark, that as already discussed in Section 2.1.3 the regularity of the Fourier transform $\hat{f}$ in the Fourier space is directly related to the decay of f for $|\vec{x}|_2 \to \infty$ in the spatial space. On this note, we introduce:

Definition 2.5. For $t + r \geq 0$, $t \geq 0$, $\hat{t} + \hat{r} \geq 0$, $\hat{t} \geq 0$ we define

$$\mathcal{H}_{\text{mix}}^{t,r;\hat{t},\hat{r}} := \left\{ f \in \mathcal{H}_{\text{mix}}^{t,r} : \hat{f} \in \mathcal{H}_{\text{mix}}^{\hat{t},\hat{r}} \subset \mathcal{L}^2 \right\} \subset \mathcal{L}^2.$$

Note that we may introduce an Hermitian inner product $\langle f, g \rangle_{\mathcal{H}_{\text{mix}}^{t,r;\hat{t},\hat{r}}} := \langle f, g \rangle_{\mathcal{H}_{\text{mix}}^{t,r}} + \langle \hat{f}, \hat{g} \rangle_{\mathcal{H}_{\text{mix}}^{\hat{t},\hat{r}}}$ with the norm $\|f\|_{\mathcal{H}_{\text{mix}}^{t,r;\hat{t},\hat{r}}}^2 := \langle f, f \rangle_{\mathcal{H}_{\text{mix}}^{t,r;\hat{t},\hat{r}}} = \|f\|_{\mathcal{H}_{\text{mix}}^{t,r}}^2 + \|\hat{f}\|_{\mathcal{H}_{\text{mix}}^{\hat{t},\hat{r}}}^2$. Now, let $\varrho_l \in C^\infty(\mathbb{R}^D)$ for all $l \in \mathbb{N}$ be given such that $\varrho_l^* \varrho_l = \eta_l$ with η_l according to (2.28) form a smooth partition of unity on $\mathbb{R}^D$. Then, corresponding to (2.29) we have with

$$\varrho_{\vec{l}} := \bigotimes_{p=1}^N \varrho_{l_p}, \quad \vec{l} \in \mathbb{N}^N,$$

2 Sobolev Spaces for Many-Particle Functions

the equivalence $\eta_{\vec{l}} = \varrho_{\vec{l}}^* \varrho_{\vec{l}}$ and hence $\{\varrho_{\vec{l}}^* \varrho_{\vec{l}}\}_{\vec{l} \in \mathbb{N}^N}$ form a smooth partition of unity on $(\mathbb{R}^D)^N$ as in (2.30). Thus, a term $f_{\vec{l}}$ given by (2.31) can be written as

$$f_{\vec{l}} = \mathcal{F}^{-1}[\varrho_{\vec{l}}^* \varrho_{\vec{l}} \hat{f}] = \mathcal{F}^{-1}[\varrho_{\vec{l}} \hat{g}_{\vec{l}}], \qquad (2.34)$$

where $\hat{g}_{\vec{l}} := \varrho_{\vec{l}}^* \hat{f}$ is at least as smooth as $\hat{f}$ and of bounded support, i.e. $\operatorname{supp} \hat{g}_{\vec{l}} \subset Q_{\vec{l}+\vec{1}}$. In particular, the support of $\hat{f}_{\mathcal{I}_L^T}$ is covered by the domain

$$\mathcal{K}_{\mathcal{I}_L^T}^{\vec{i}} := \bigcup_{\vec{l} \in \mathcal{T}_L^T} Q_{\vec{l}+\vec{1}}.$$

To obtain a finite-dimensional general sparse grid space each $\hat{g}_{\vec{l}}$, $\vec{l} \in \mathcal{I}_L^T$ may now be approximated by a spectral Galerkin method. To this end, let us suppose that we have an orthonormal basis $\{b_{\mathbf{j}}\}_{\mathbf{j} \in \mathbb{Z}^D}$ for the space of square integrable functions on a D-dimensional torus $\mathbb{T}_a^D$, $\mathbb{T}_a := [-\frac{a}{2}, +\frac{a}{2}]$, $a > 0$, $\mathcal{L}^2(\mathbb{T}_a^D)$. Let us remark that, except for the completion with respect to a chosen Sobolev norm, the union $\bigcup_{J=1}^{\infty} V_J$ with $V_J := \operatorname{span}\{b_{\mathbf{j}} : \mathbf{j} \in \mathbb{Z}^D, |\mathbf{j}|_\infty \leq 2^J\}$ is just the associated Sobolev space. Furthermore, we introduce the functions

$$b_{l,\mathbf{j}}(\mathbf{k}) := (a_{l+1})^{-\frac{D}{2}} b_{\mathbf{j}}\left(\frac{1}{a_{l+1}} \mathbf{k}\right) \qquad (2.35)$$

with $a_l := \frac{2}{a}(2^{2l} - 1)^{\frac{1}{2}}$, which form an orthonormal basis for $\mathcal{L}^2(Q_{l+1})$ and with it the functions

$$b_{\vec{l},\vec{\mathbf{j}}}(\vec{\mathbf{k}}) := \bigotimes_{p=1}^{N} b_{l_p, \mathbf{j}_p}(\vec{\mathbf{k}}) = |\det A_{\vec{l}+\vec{1}}|^{-\frac{1}{2}} \bigotimes_{p=1}^{N} b_{\mathbf{j}_p}(A_{\vec{l}+\vec{1}}^{-1} \vec{\mathbf{k}}), \quad \vec{\mathbf{j}} \in (\mathbb{Z}^D)^N,$$

which then form an orthonormal basis for $\mathcal{L}^2(Q_{\vec{l}+\vec{1}})$, where the linear transformation $A_{\vec{l}}$ is given by

$$A_{\vec{l}} : (\mathbb{T}_a^D)^N \to Q_{\vec{l}} : \vec{\mathbf{k}} \mapsto (a_{l_1} \mathbf{k}_1, \ldots, a_{l_N} \mathbf{k}_N). \qquad (2.36)$$

We can then uniquely represent any $\hat{g}_{\vec{l}}$ as

$$\hat{g}_{\vec{l}}(\vec{\mathbf{k}}) = \sum_{\vec{\mathbf{j}} \in (\mathbb{Z}^D)^N} c_{\vec{l},\vec{\mathbf{j}}} b_{\vec{l},\vec{\mathbf{j}}}(\vec{\mathbf{k}}) \qquad (2.37)$$

with coefficients

$$c_{\vec{l},\vec{\mathbf{j}}} = \int_{Q_{\vec{l}+\vec{1}}} b_{\vec{l},\vec{\mathbf{j}}}^*(\vec{\mathbf{k}}) \hat{g}_{\vec{l}}(\vec{\mathbf{k}}) \, d\vec{\mathbf{k}} = \langle b_{\vec{l},\vec{\mathbf{j}}}, \hat{g}_{\vec{l}} \rangle_{\mathcal{L}^2} \qquad (2.38)$$

and we especially have $\|\hat{g}_{\vec{l}}\|_{\mathcal{L}^2((\mathbb{R}^D)^N)}^2 = \|\hat{g}_{\vec{l}}\|_{\mathcal{L}^2(Q_{\vec{l}+\vec{1}})}^2 = \sum_{\vec{\mathbf{j}} \in (\mathbb{Z}^D)^N} |c_{\vec{l},\vec{\mathbf{j}}}|^2 = \|\{c_{\vec{l},\vec{\mathbf{j}}}\}_{\vec{\mathbf{j}}}\|_{\ell^2}^2$. Note that the coefficients given in (2.38) may be also written with the help of the functions

$$\phi_{\vec{l},\vec{\mathbf{j}}} := \mathcal{F}^{-1}[\varrho_{\vec{l}} b_{\vec{l},\vec{\mathbf{j}}}] = \bigotimes_{p=1}^{N} \phi_{l_p, \mathbf{j}_p}, \quad \phi_{l,\mathbf{j}}(\mathbf{x}) := \mathcal{F}^{-1}[\varrho_l b_{l,\mathbf{j}}](\mathbf{x}) \qquad (2.39)$$

in the form

$$c_{\vec{l},\vec{\mathbf{j}}} = \int_{Q_{\vec{l}+\vec{1}}} (\varrho_{\vec{l}}(\vec{\mathbf{k}}) b_{\vec{l},\vec{\mathbf{j}}}(\vec{\mathbf{k}}))^* \hat{f}(\vec{\mathbf{k}}) \, d\vec{\mathbf{k}} = \langle \phi_{\vec{l},\vec{\mathbf{j}}}, f \rangle_{\mathcal{L}^2((\mathbb{R}^D)^N)}.$$

2.3 Hyperbolic cross approximation

Inserting the series representation (2.37) in (2.34) results in

$$f_{\vec{l}} = \mathcal{F}^{-1}[\varrho_{\vec{l}} \sum_{\vec{j} \in (\mathbb{Z}^D)^N} c_{\vec{l}\vec{j}} b_{\vec{l}\vec{j}}]. \tag{2.40}$$

See also [45, 57, 139]. Now, if we denote the partial sum of the series representation (2.40) associated with a finite subset $\mathcal{J}_J^R(\vec{l}) \subset (\mathbb{Z}^D)^N$ by

$$f_{\vec{l},\mathcal{J}_J^R(\vec{l})} := \mathcal{F}^{-1}[\varrho_{\vec{l}} \sum_{\vec{j} \in \mathcal{J}_J^R(\vec{l})} c_{\vec{l}\vec{j}} b_{\vec{l}\vec{j}}]$$

and introduce

$$f_{\mathcal{I}_L^T, \mathcal{J}_J^R}(\vec{x}) := \sum_{\vec{l} \in \mathcal{I}_L^T} f_{\vec{l},\mathcal{J}_J^R(\vec{l})}, \tag{2.41}$$

then it holds the relation

$$\|f - f_{\mathcal{I}_L^T, \mathcal{J}_J^R}\|_{\mathcal{H}_{\text{mix}}^{t',r'}} = \|f - f_{\mathcal{I}_L^T} + f_{\mathcal{I}_L^T} - f_{\mathcal{I}_L^T, \mathcal{J}_J^R}\|_{\mathcal{H}_{\text{mix}}^{t',r'}} \\ \leq \|f - f_{\mathcal{I}_L^T}\|_{\mathcal{H}_{\text{mix}}^{t',r'}} + \|f_{\mathcal{I}_L^T} - f_{\mathcal{I}_L^T, \mathcal{J}_J^R}\|_{\mathcal{H}_{\text{mix}}^{t',r'}} \tag{2.42}$$

for $t' + r' \leq t + r$, $t - t' \geq 0$. Furthermore, for either $t' = 0$ or $t' > 0$, $\frac{r'}{t'} \geq T$ the estimate

$$\|f_{\mathcal{I}_L^T} - f_{\mathcal{I}_L^T, \mathcal{J}_J^R}\|_{\mathcal{H}_{\text{mix}}^{t',r'}}^2 = \int_{(\mathbb{R}^D)^N} (w_{\text{mix}}(\vec{k}))^{2t'} (w_{\text{iso}}(\vec{k}))^{2r'} |\hat{f}_{\mathcal{I}_L^T}(\vec{k}) - \hat{f}_{\mathcal{I}_L^T, \mathcal{J}_J^R}(\vec{k})|_2^2 \, d\vec{k} \\ \leq \left(\sup_{\vec{k} \in \mathcal{K}_{\mathcal{I}_L^T}^T} (w_{\text{mix}}(\vec{k}))^{2t'} (w_{\text{iso}}(\vec{k}))^{2r'} \right) \int_{\mathcal{K}_{\mathcal{I}_L^T}^T} |\hat{f}_{\mathcal{I}_L^T}(\vec{k}) - \hat{f}_{\mathcal{I}_L^T, \mathcal{J}_J^R}(\vec{k})|_2^2 \, d\vec{k} \\ \lesssim 2^{2L(t'+r')} \|\hat{f}_{\mathcal{I}_L^T} - \hat{f}_{\mathcal{I}_L^T, \mathcal{J}_J^R}\|_{\mathcal{L}^2}^2 \tag{2.43}$$

holds. In addition,

$$\|\hat{f}_{\mathcal{I}_L^T} - \hat{f}_{\mathcal{I}_L^T, \mathcal{J}_J^R}\|_{\mathcal{L}^2} = \|\sum_{\vec{l} \in \mathcal{I}_L^T} (\hat{f}_{\vec{l}} - \hat{f}_{\vec{l},\mathcal{J}_J^R(\vec{l})})\|_{\mathcal{L}^2} \leq \sum_{\vec{l} \in \mathcal{I}_L^T} \|\hat{f}_{\vec{l}} - \hat{f}_{\vec{l},\mathcal{J}_J^R(\vec{l})}\|_{\mathcal{L}^2} \tag{2.44}$$

and for the approximation error with respect to the $\mathcal{L}^2$-norm we obtain the estimate

$$\|\hat{f}_{\vec{l}} - \hat{f}_{\vec{l},\mathcal{J}_J^R(\vec{l})}\|_{\mathcal{L}^2}^2 = \|\varrho_{\vec{l}} \sum_{\vec{j} \in \overline{\mathcal{J}_J^R(\vec{l})}} c_{\vec{l}\vec{j}} b_{\vec{l}\vec{j}}\|_{\mathcal{L}^2}^2 \leq \|\sum_{\vec{j} \in \overline{\mathcal{J}_J^R(\vec{l})}} c_{\vec{l}\vec{j}} b_{\vec{l}\vec{j}}\|_{\mathcal{L}^2}^2 = \sum_{\vec{j} \in \overline{\mathcal{J}_J^R(\vec{l})}} |c_{\vec{l}\vec{j}}|^2, \tag{2.45}$$

where $\overline{\mathcal{J}_J^R(\vec{l})} := (\mathbb{Z}^D)^N \setminus \mathcal{J}_J^R(\vec{l})$. Now, let us assume that the system $\{b_{\vec{j}}\}_{\vec{j} \in \mathbb{Z}^D}$ and the sequence of sets $\{\mathcal{J}_J^R(\vec{l})\}_{J \in \mathbb{N}}$ are chosen such that each part $f_{\vec{l}}$ can be approximated by appropriate partial sums $f_{\vec{l},\mathcal{J}_J^R(\vec{l})}$ up to arbitrary accuracy with respect to the $\mathcal{L}^2$-norm in the space

$$V_{\vec{l}} := \text{span}\{\phi_{\vec{l}\vec{j}} : \vec{j} \in (\mathbb{Z}^D)^N\},$$

i.e. the sum on the right hand side of (2.45) can be made arbitrary small by an increase of the parameter J. Here, $\phi_{\vec{l}\vec{j}}$ is defined according to (2.39). Then, due to (2.44) and (2.43), the band-limited functions $f_{\mathcal{I}_L^T}$ can also arbitrarily well be approximated by appropriate partial sums $f_{\mathcal{I}_L^T,\mathcal{J}_j^R}$ in the space $V_L^T := \sum_{\vec{l}\in\mathcal{I}_L^T} V_{\vec{l}}$ with respect to the $\mathcal{H}_{\text{mix}}^{t',r'}$-norm. Hence, both terms on the right hand side of (2.42) can be made arbitrarily small and thus a function $f \in \mathcal{H}_{\text{mix}}^{t,r;\hat{t},\hat{r}}$ according to Definition 2.5 can also be approximated arbitrarily well by appropriate partial sums $f_{\mathcal{I}_L^T,\mathcal{J}_j^R}$ with respect to the $\mathcal{H}_{\text{mix}}^{t',r'}$-norm. Note that such a space V_L^T can be associated with an infinitely extended general sparse grid. Let us remark here that it is easy to adapt the latter construction to the case of using a partition of unity $\{\eta_{\vec{l}}\}_{\vec{l}\in\mathbb{N}^N}$ corresponding to (2.30) with $\eta_{\vec{l}} = \tilde{\varrho}_{\vec{l}}^* \varrho_{\vec{l}}$ where $\tilde{\varrho}_{\vec{l}}, \varrho_{\vec{l}} \in C^\infty$ and also of using biorthonormal Riesz bases $\{\tilde{b}_{\mathbf{j}}\}_{\mathbf{j}\in\mathbb{Z}^D}$, $\{b_{\mathbf{j}}\}_{\mathbf{j}\in\mathbb{Z}^D}$ instead of an orthonormal basis. Especially, we would then have $f_{\vec{l}} = \sum_{\vec{\mathbf{j}}\in\mathcal{J}_j^R(\vec{l})} c_{\vec{l}\vec{\mathbf{j}}} \mathcal{F}^{-1}[\varrho_{\vec{l}} b_{\vec{l}\vec{\mathbf{j}}}]$ and $\|f_{\vec{l}} - f_{\vec{l},\mathcal{J}_j^R(\vec{l})}\|_{\mathcal{L}^2}^2 \lesssim \sum_{\vec{\mathbf{j}}\in\overline{\mathcal{J}_j^R(\vec{l})}} |c_{\vec{l}\vec{\mathbf{j}}}|^2$, where the coefficients are given by $c_{\vec{l}\vec{\mathbf{j}}} = \int (\tilde{\varrho}_{\vec{l}}(\vec{\mathbf{k}})\tilde{b}_{\vec{l}\vec{\mathbf{j}}}(\vec{\mathbf{k}}))^* \hat{f}(\vec{\mathbf{k}}) \, d\vec{\mathbf{k}}$. For a further reading on using biorthonormal Riesz bases in particular with regard to stability of tensor product bases and norm equivalences for spaces built from intersections of tensor products of Hilbert spaces, e.g. $\mathcal{H}_{\text{mix}}^{t,l}$ with $l \leq 0$, spaces see also [72, 73]. In the following we discuss the use of an orthonormal Riesz basis in more detail, exemplified by an expansion of each $\hat{g}_{\vec{l}}$ in a Fourier series and using a smooth partition of unity with $\eta_{\vec{l}} = \varrho_{\vec{l}}^* \varrho_{\vec{l}}$.

Exemplary use of a Fourier series expansion

To this end, we employ for the basis $\{b_{\mathbf{j}}\}_{\mathbf{j}\in\mathbb{Z}^D}$ in (2.35) the Fourier basis $\tilde{e}_{\mathbf{j}}$, $\mathbf{j} \in \mathbb{Z}^D$ according to (2.18), which especially is an orthonormal basis of $\mathcal{L}^2(\mathbb{T}^D)$, where $\mathbb{T}^D = \mathbb{T}_a^D$, $a = 2\pi$. Moreover, for reasons of simplicity we set $Q_{\vec{l}}$ to $\tilde{Q}_{\vec{l}}$ from (2.27) and with it we set $a_l := \frac{2}{a} 2^l$ for the linear transform in (2.36). In particular, the functions $\phi_{\vec{l}\vec{\mathbf{j}}}$ in (2.39) can then be written in the form

$$\phi_{\vec{l}\vec{\mathbf{j}}}(\vec{\mathbf{x}}) = \bigotimes_{p=1}^N \phi_{l_p,\mathbf{j}_p}(\vec{\mathbf{x}}) = \bigotimes_{p=1}^N \phi_{l_p}(\vec{\mathbf{x}} + A_{\vec{l}+\vec{1}}^{-1}\vec{\mathbf{j}}), \quad \phi_l := (a_{l+1} 2\pi)^{-\frac{D}{2}} \mathcal{F}^{-1}[\varrho_l] \qquad (2.46)$$

and for a certain resolution of unity $\{\varrho_{\vec{l}}^* \varrho_{\vec{l}}\}_{\vec{l}\in\mathbb{N}^N}$ correspond to an orthonormal tensor product basis built from Meyer wavelets and Meyer scaling functions; see Appendix B.2. Also, we assume that $\hat{t}, \hat{r} \in \mathbb{N}_0$ and that there exist $C_{\vec{\mathbf{m}}} > 0$ for all $\vec{\mathbf{m}} \in (\mathbb{N}_0^D)^N$, such that for all $\vec{l} \in \mathbb{N}^N$ the estimate

$$D_{\vec{\mathbf{k}}}^{\vec{\mathbf{m}}} \varrho_{\vec{l}}(\vec{\mathbf{k}}) \leq C_{\vec{\mathbf{m}}} \varrho_{\vec{l}}(\vec{\mathbf{k}}) \quad \text{for all } \vec{\mathbf{k}} \in (\mathbb{R}^D)^N, \qquad (2.47)$$

holds.

Let $u_{\vec{l}}$ be a square integrable periodic function on $Q_{\vec{l}+\vec{1}}$, with the unique representation $u_{\vec{l}}(\vec{\mathbf{k}}) = \sum_{\vec{\mathbf{j}}\in(\mathbb{Z}^D)^N} u_{\vec{l}\vec{\mathbf{j}}} b_{\vec{l}\vec{\mathbf{j}}}(\vec{\mathbf{k}})$. Here, we have for the $\mathcal{H}_{\text{mix}}^{\hat{t},\hat{r}}$-norm as in (2.12) the equivalence

$$\|u_{\vec{l}}\|_{\mathcal{H}_{\text{mix}}^{\hat{t},\hat{r}}(Q_{\vec{l}+\vec{1}})}^2 \simeq \sum_{\vec{\mathbf{j}}\in(\mathbb{Z}^D)^N} (w_{\text{mix}}(A_{\vec{l}+\vec{1}}^{-1}\vec{\mathbf{j}}))^{2\hat{t}} (w_{\text{iso}}(A_{\vec{l}+\vec{1}}^{-1}\vec{\mathbf{j}}))^{2\hat{r}} |u_{\vec{l}\vec{\mathbf{j}}}|^2,$$

where terms are included on the right hand side which also depend on $\vec{l}$. Now, let $\tilde{\hat{g}}$ be the periodic extension of $\hat{g} \in \mathcal{H}_{\text{mix}}^{\hat{t},\hat{r}}(Q_{\vec{l}+\vec{1}})$ onto $(\mathbb{R}^D)^N$. Then, it holds the relation

$$\sum_{\vec{\mathbf{j}}\in\mathcal{J}_j^R(\vec{l})} (w_{\text{mix}}(A_{\vec{l}+\vec{1}}^{-1}\vec{\mathbf{j}}))^{2\hat{t}} (w_{\text{iso}}(A_{\vec{l}+\vec{1}}^{-1}\vec{\mathbf{j}}))^{2\hat{r}} |c_{\vec{l}\vec{\mathbf{j}}}|^2 \simeq \|\tilde{\hat{g}}_{\vec{l}}\|_{\mathcal{H}_{\text{mix}}^{\hat{t},\hat{r}}(Q_{\vec{l}+\vec{1}})}^2 = \|\hat{g}_{\vec{l}}\|_{\mathcal{H}_{\text{mix}}^{\hat{t},\hat{r}}((\mathbb{R}^D)^N)}^2.$$

Furthermore, we define the subsets $\mathcal{J}_J^R(\vec{l}) \subset (\mathbb{Z}^D)^N$ by

$$\mathcal{J}_J^R(\vec{l}) := \bigcup_{\vec{\iota} \in \mathcal{I}_J^R} \mathcal{I}_{\vec{\iota}+\vec{l}+\vec{1}}, \quad \mathcal{I}_{\vec{\alpha}} := \mathcal{I}_{\alpha_1} \times \cdots \times \mathcal{I}_{\alpha_N}, \quad \mathcal{I}_{\alpha} := \mathbb{Z}^D \cap Q_\alpha, \qquad (2.48)$$

i.e. $\mathcal{I}_\alpha \subset \{\mathbf{k} \in \mathbb{Z}^D : |\mathbf{k}|_\infty \leq 2^\alpha\}$. This yields the relation

$$\sum_{\vec{j} \in \mathcal{J}_J^R(\vec{l})} |c_{\vec{l}\vec{j}}|^2 = \sum_{\vec{j} \in \mathcal{J}_J^R(\vec{l})} \frac{(w_{\text{iso}}(A_{\vec{l}+\vec{1}}^{-1}\vec{j}))^{-2\hat{r}}}{(w_{\text{mix}}(A_{\vec{l}+\vec{1}}^{-1}\vec{j}))^{2\hat{t}}} (w_{\text{mix}}(A_{\vec{l}+\vec{1}}^{-1}\vec{j}))^{2\hat{t}} (w_{\text{iso}}(A_{\vec{l}+\vec{1}}^{-1}\vec{j}))^{2\hat{r}} |c_{\vec{l}\vec{j}}|^2$$

$$\leq \left(\max_{\vec{j} \in \mathcal{J}_J^R(\vec{l})} \frac{(w_{\text{iso}}(A_{\vec{l}+\vec{1}}^{-1}\vec{j}))^{-2\hat{r}}}{(w_{\text{mix}}(A_{\vec{l}+\vec{1}}^{-1}\vec{j}))^{2\hat{t}}} \right) \sum_{\vec{j} \in \mathcal{J}_J^R(\vec{l})} (w_{\text{mix}}(A_{\vec{l}+\vec{1}}^{-1}\vec{j}))^{2\hat{t}} (w_{\text{iso}}(A_{\vec{l}+\vec{1}}^{-1}\vec{j}))^{2\hat{r}} |c_{\vec{l}\vec{j}}|^2$$

$$\simeq \left(\max_{\vec{\iota} \in \mathbb{N}^N \setminus \mathcal{I}_J^R} 2^{-2\hat{t}|\vec{\iota}|_1 - 2\hat{r}|\vec{\iota}|_\infty} \right) \|\hat{g}_{\vec{l}}\|_{\mathcal{H}_{\text{mix}}^{\hat{t},\hat{r}}((\mathbb{R}^D)^N)}^2 \qquad (2.49)$$

for the right hand side of (2.45). We summarize the discussion in a lemma.

Lemma 2.5. *Let $r' + t' \geq 0$, $t' \geq 0$, $t' + r' < t + r$, $t - t' \geq 0$, $\hat{r}, \hat{t} \in \mathbb{N}_0$, $\hat{t} + \hat{r} > 0$, $f \in \mathcal{H}_{\text{mix}}^{t,r;\hat{t},\hat{r}}((\mathbb{R}^D)^N)$, $f_{\mathcal{I}_L^T}$ as in (2.33) and $f_{\mathcal{I}_L^T, \mathcal{J}_J^R}$ as in (2.41) (where $\mathbf{b_j}$, $\mathbf{j} \in \mathbb{Z}^D$ is set to the Fourier basis $\tilde{e}_\mathbf{j}$, $\mathbf{j} \in \mathbb{Z}^D$ according to (2.18)). Additionally, let either $t' = 0$ or $t' > 0$, $\frac{r'}{t'} \geq T$. Then for $\hat{t} > 0$ it holds that*

$$\|f_{\mathcal{I}_L^T} - f_{\mathcal{I}_L^T, \mathcal{J}_J^R}\|_{\mathcal{H}_{\text{mix}}^{t',r'}} \lesssim \begin{cases} 2^{L(t'+r')} 2^{-J(\hat{t}+\hat{r}-(R\hat{t}+\hat{r})\frac{N-1}{N-R})} \|\hat{f}\|_{\mathcal{H}_{\text{mix}}^{\hat{t},\hat{r}}} & \text{for } R \geq -\frac{\hat{r}}{\hat{t}}, \\ 2^{L(t'+r')} 2^{-J(\hat{t}+\hat{r})} \|\hat{f}\|_{\mathcal{H}_{\text{mix}}^{\hat{t},\hat{r}}} & \text{for } R \leq -\frac{\hat{r}}{\hat{t}}, \end{cases}$$

and for $\hat{t} = 0$ it holds that

$$\|f_{\mathcal{I}_L^T} - f_{\mathcal{I}_L^T, \mathcal{J}_J^R}\|_{\mathcal{H}_{\text{mix}}^{t',r'}} \lesssim \begin{cases} 2^{L(t'+r')} 2^{-J\hat{r}(1-\frac{N-1}{N-R})} \|\hat{f}\|_{\mathcal{H}_{\text{mix}}^{\hat{t},\hat{r}}} & \text{for } R > -\infty, \\ 2^{L(t'+r')} 2^{-J\hat{r}} \|\hat{f}\|_{\mathcal{H}_{\text{mix}}^{\hat{t},\hat{r}}} & \text{for } R = -\infty. \end{cases}$$

Proof. Evaluating the maximum in (2.49) leads to the desired result with (2.43), (2.44), (2.45) and (2.47). □

Here, the function $f_{\mathcal{I}_L^T, \mathcal{J}_J^R}$ is in a finite-dimensional space

$$V_{L;J}^{T;R} := \text{span}(\mathcal{B}_{V_{L;J}^{T;R}}), \quad \mathcal{B}_{V_{L;J}^{T;R}} := \{\phi_{\vec{l}\vec{j}} : \vec{l} \in \mathcal{I}_L^T, \vec{j} \in \mathcal{J}_J^R(\vec{l})\}, \qquad (2.50)$$

which is a subspace of V_L^T and can be particularly associated with a general sparse grid.[3] Examples for the general sparse grid spaces $V_{L;J}^{T;R}$ in the case of the Meyer wavelet family with $D = 1$ and $N = 2$ are shown in Figure 2.3 and in Figure 2.4.[4] Obviously, under the

---
[3] Let us remark that, except for the completion with respect to a chosen Sobolev norm, the union $\bigcup_{L=1}^\infty \bigcup_{J=1}^\infty V_{L;J}^{T;R}$ is just the associated Sobolev space.
[4] Note that for the adaption to the case of the Meyer wavelet family given in B.2, we slightly modify the definition of index sets $\mathcal{J}_J^R(\vec{l})$, where the approximation and complexity orders are not changed.

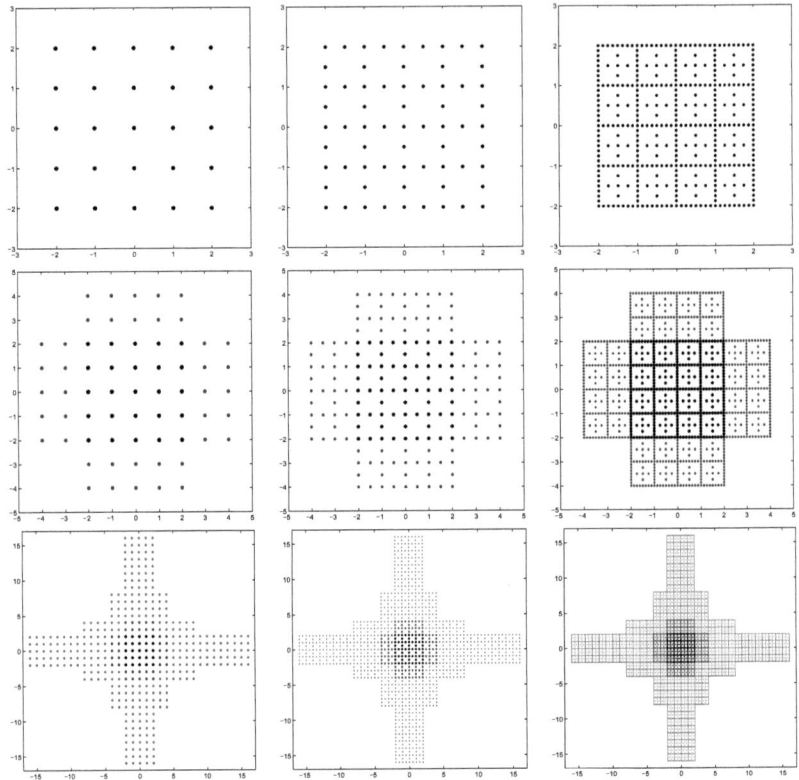

Figure 2.3: Localization peaks of basis functions in $V_{L;J}^{0;0}$ according to (2.50) with $D = 1$ and $N = 2$. Here, we have $L = 1, 2, 4$ from left to right and $J = 1, 2, 4$ from top to bottom.

assumptions of Lemma 2.5, an upper bound estimate for the best approximation error of $f \in \mathcal{H}_{\text{mix}}^{t,r;\hat{t},\hat{r}}$ in the $\mathcal{H}_{\text{mix}}^{t',r'}$-norm is given with

$$\inf_{\tilde{f} \in V_{L;J}^{T;R}} \|f - \tilde{f}\|_{\mathcal{H}_{\text{mix}}^{t',r'}} \leq \|f - f_{\mathcal{I}_L^T, \mathcal{J}_J^R}\|_{\mathcal{H}_{\text{mix}}^{t',r'}} \qquad (2.51)$$

by equation (2.42) together with Lemma 2.4 and Lemma 2.5. Here, the aim is to choose the parameters such that the two error terms given in Lemma 2.4 and Lemma 2.5 are balanced. Let us shortly discuss the case of $f \in \mathcal{H}_{\text{mix}}^{t,r;\hat{t},\hat{r}}$ with the assumptions of Lemma 2.5 and $T \leq \frac{r'-r}{t-t'}$, $R \leq -\frac{\hat{r}}{\hat{t}}$. Then it holds the estimate

$$\inf_{\tilde{f} \in V_{L;J}^{T;R}} \|f - \tilde{f}\|_{\mathcal{H}_{\text{mix}}^{t',r'}} \lesssim 2^{L((t'+r')-(t+r))} \|f\|_{\mathcal{H}_{\text{mix}}^{t,r}} + 2^{L\left((t'+r')-\frac{J}{L}(\hat{t}+\hat{r})\right)} \|\hat{f}\|_{\mathcal{H}_{\text{mix}}^{\hat{t},\hat{r}}}$$

2.3 Hyperbolic cross approximation

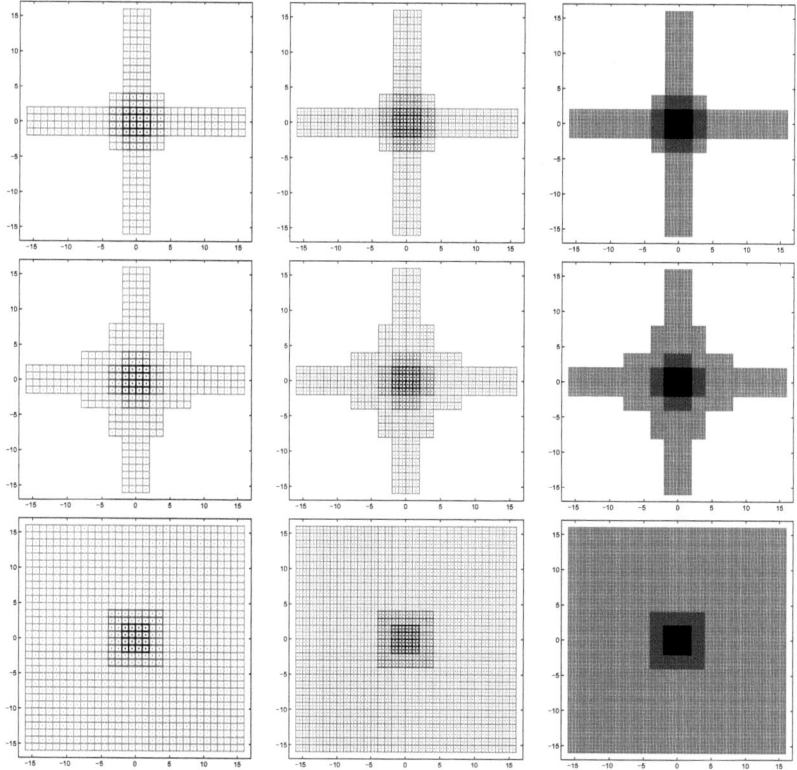

Figure 2.4: Localization peaks of basis functions in $V_{4;4}^{T;R}$ according to (2.50) with $D = 1$ and $N = 2$. Here, we have $T = \frac{1}{2}, 0, -\infty$ from left to right and $R = \frac{1}{2}, 0, -\infty$ from top to bottom.

and for $\tilde{J}(L) := \left\lceil L\frac{r+t}{\hat{t}+\hat{r}} \right\rceil$ in particular

$$L(r+t) \leq \tilde{J}(L)(\hat{t}+\hat{r}). \tag{2.52}$$

Thus, we obtain the relation

$$\inf_{\tilde{f} \in V_{L;\tilde{J}(L)}^{T;R}} \|f - \tilde{f}\|_{\mathcal{H}_{\text{mix}}^{t',r'}} \lesssim 2^{L(t'+r'-(t+r))} \left(\|f\|_{\mathcal{H}_{\text{mix}}^{t,r}} + \|\hat{f}\|_{\mathcal{H}_{\text{mix}}^{\hat{t},\hat{r}}} \right). \tag{2.53}$$

In this case, the approximation rate according to Lemma 2.4 is preserved.[5]

---

[5]We remark that for $\hat{f} \in C^{\infty}$ the parameters $\hat{r}$ and $\hat{t}$ can be chosen arbitrarily high in dependence of L such that the relation $(t+r) \leq \frac{J}{L}(\hat{t}+\hat{r})$ is fulfilled for a fixed J. However, the constants involved in the estimates are then also dependent on $\hat{r}$ and $\hat{t}$.

The number of degrees of freedom, i.e. the dimension of $V_{L;J}^{T;R}$, is equal to $|\mathcal{B}_{V_{L;J}^{T;R}}|$ and thus to the cardinality number of the index set

$$\mathcal{I}_{L;J}^{T;R} := \left\{ (\vec{l}, \vec{j}) : \vec{l} \in \mathcal{I}_L^T, \vec{j} \in \mathcal{J}_L^R(\vec{l}) \right\} \subset \mathbb{N}^N \times (\mathbb{Z}^D)^N. \tag{2.54}$$

Let us introduce the index set $\mathcal{I}_L^{T;(D)} := \bigcup_{\vec{l} \in \mathcal{I}_L^T} \mathcal{I}_{\vec{l}}$ with $\mathcal{I}_{\vec{l}}$ from (2.48), which leads to a generalization of Lemma 2.2 by:

Lemma 2.6. *For $T < 1$, $L \in \mathbb{N}$ it holds*

$$|\mathcal{I}_L^{T;(D)}| \lesssim \begin{cases} \mathcal{O}(2^{DL}) & \text{for } 0 < T < 1, \\ \mathcal{O}(2^{DL} L^{N-1}) & \text{for } T = 0, \\ \mathcal{O}(2^{DL \frac{T-1}{T/N-1}}) & \text{for } T < 0, \\ \mathcal{O}(2^{DLN}) & \text{for } T = -\infty. \end{cases} \tag{2.55}$$

Proof. Since it holds

$$|\mathcal{I}_L^{T;(D)}| \lesssim \sum_{\vec{l} \in \mathcal{I}_L^T} 2^{D|\vec{l}|_1} \lesssim \Big(\sum_{\vec{l} \in \mathcal{I}_L^T} 2^{|\vec{l}|_1 - N} \Big)^D$$

estimate (2.55) follows directly with Lemma 2.2 for all cases except for $T = 0$. The case of $T = 0$ results from

$$\sum_{\vec{l} \in \mathcal{I}_L^0} 2^{D(|\vec{l}|_1 - N)} = \sum_{j=N}^{L+N-1} 2^{D(j-N)} \sum_{|\vec{l}|=j} 1$$

$$= \sum_{j=0}^{L-1} 2^{Dj} \binom{N-1+j}{N-1}$$

$$= \frac{1}{(N-1)!} \sum_{j=0}^{L-1} \left(x^{j+N-1} \right)^{(N-1)} \Big|_{x=2^D}$$

$$= \frac{1}{(N-1)!} \left(x^{N-1} \frac{1-x^L}{1-x} \right)^{(N-1)} \Big|_{x=2^D}$$

$$= \frac{1}{(N-1)!} \sum_{j=0}^{N-1} \binom{N-1}{j} \left(x^{N-1} - x^{L+N-1} \right)^{(j)} \left(\frac{1}{1-x} \right)^{(N-1-j)} \Big|_{x=2^D}$$

$$\lesssim 2^{DL} \sum_{j=0}^{N-1} \binom{L+N-1}{j} 2^{D(N-1-j)}$$

$$\lesssim 2^{DL} L^{N-1}.$$

See also [24, 73] and selected references therein. □

An upper estimate for $|\mathcal{I}_{L;J}^{T;R}|$ can given with the help of Lemma 2.6 and the following lemma:

Lemma 2.7. *For $T < 1$, $L \in \mathbb{N}$, $R < 1$, $J \in \mathbb{N}$ it holds*

$$|\mathcal{I}_{L;J}^{T;R}| \lesssim |\mathcal{I}_L^{T;(D)}| \, |\mathcal{I}_J^{R;(D)}|. \tag{2.56}$$

2.3 Hyperbolic cross approximation

Proof. We have $|\mathcal{I}_{\vec{\alpha}}| \lesssim 2^{D|\vec{\alpha}|_1}$ and thus with (2.54) and (2.48), relation (2.56) easily follows with

$$|\mathcal{I}_{L;J}^{T;R}| \lesssim \sum_{\vec{l} \in \mathcal{I}_L^T} \sum_{\vec{i} \in \mathcal{J}_J^R} 2^{D|\vec{i}+\vec{l}+\vec{1}|_1}$$

$$\lesssim \sum_{\vec{l} \in \mathcal{I}_L^T} 2^{D|\vec{l}|_1} \sum_{\vec{i} \in \mathcal{I}_J^R} 2^{D|\vec{i}|_1}.$$

□

All in all, according to estimate (2.53) and Lemma 2.7 the following result holds: For functions in $\mathcal{H}_{\text{mix}}^{t,r;\hat{t},\hat{r}}$ with the assumptions of Lemma 2.5 the use of the generalized sparse grid space $V_{L;\tilde{J}(L)}^{T;R}$ with $T \leq \frac{r'-r}{t-t'}$, $R \leq -\frac{\hat{r}}{\hat{t}}$ and (2.52) leads to a significant reduction in the number of degrees of freedom compared to the full grid space $V_{L;\tilde{J}(L)}^{-\infty;R}$, while the approximation order is preserved.

Let us discuss some cases. For the spaces $\mathcal{H}_{\text{mix}}^{0,r;\hat{t},\hat{r}}$ (i.e. $t = t' = 0$, $0 \leq r' < r$, $\hat{t} + \hat{r} > 0$) and the spaces $V_{L;\tilde{J}(L)}^{T;R}$ (i.e. $R \leq -\frac{\hat{r}}{\hat{t}}$) with $T > -\infty$ the resulting approximation order with respect to L is dependent on T and dependent on the number of particles N. In particular the order even deteriorates with larger T, whereas for $T = -\infty$ the order is independent on the number of particles N. However, since for $T < 0$ the dimension of $\mathcal{I}_L^{T;(D)}$ with respect to L is exponentially dependent on N, the dimension of $V_{L;\tilde{J}(L)}^{T;R}$ with respect to L is also exponentially dependent on N. This reflects the curse of dimensionality which makes problems in isotropic Sobolev spaces $\mathcal{H}_{\text{mix}}^{0,r}$ intractable for higher values of N. For the spaces of bounded mixed derivatives $\mathcal{H}_{\text{mix}}^{t,0;\hat{t},\hat{r}}$ (i.e. $r = t' = 0$, $0 \leq r' < t$, $\hat{t} + \hat{r} > 0$) and the spaces $V_{L;\tilde{J}(L)}^{T;R}$ (i.e. $R \leq -\frac{\hat{r}}{\hat{t}}$) with $T > \frac{r'}{t}$ the resulting approximation order is dependent on T and dependent on the number of particles N whereas for $T \leq \frac{r'}{t}$ the resulting order is independent of T and N. Note that for $T > 0$ the dimension of $\mathcal{I}_L^{T;(D)}$ with respect to L is independent of N. Thus, for $T \in (0, \frac{r'}{t}]$ the dependency of the dimension of $V_{L;\tilde{J}(L)}^{T;R}$ on N is given by the factor $|\mathcal{I}_{\tilde{J}(L)}^{R;(D)}|$ only. For example if we restrict the class of functions to $\mathcal{H}_{\text{mix}}^{t,r;\hat{t},\hat{r}}$ (i.e. $t > 0$, $r \geq 0$, $\hat{t} + \hat{r} > 0$, $R \leq -\frac{\hat{r}}{\hat{t}}$) and measure the error in the $\mathcal{H}^r$-norm (i.e. $t' = 0$, $r' = r$), the approximation order with respect to L is dependent on N for all $T > 0$ and independent on N and T for all $T \leq 0$. In that case, for $T = 0$, the dependence of the cardinality number $|\mathcal{I}_L^{T;(D)}|$ on N is only logarithmic. This reflects the fact that it is here possible to get rid of the curse of dimensionality of the discretization of the standard isotropic Sobolev spaces at least to some extent. Note that in all cases the constants in the $\mathcal{O}$-notation depend on N and D.

We cast the estimates on the degrees of freedom and the associated error into a form which measures the error with respect to the involved degrees of freedom and reach the following lemma in a special case:

Lemma 2.8. *Let* $t' + r' \geq 0$, $t' \geq 0$, $t' + r' < t + r$, $t - t' \geq 0$, $\hat{t} \in \mathbb{N}_0$, $\hat{t} > 0$, $f \in \mathcal{H}_{\text{mix}}^{t,r;\hat{t},0}$, $0 = T \leq \frac{r'-r}{t-t'}$, $R = 0$ *and* $\tilde{J}(L) = \frac{t+r}{\hat{t}} L$. *Then with* $M := |V_{L;\tilde{J}(L)}^{T;R}|$ *it holds*

$$\inf_{\tilde{f} \in V_{L;\tilde{J}(L)}^{T;R}} \|f - \tilde{f}\|_{\mathcal{H}_{\text{mix}}^{t',r'}} \lesssim \left(\frac{M}{\log_2(M)^{2(N-1)}}\right)^{-\frac{(t+r-(t'+r'))}{D(1+(t+r)/\hat{t})}} \|f\|_{\mathcal{H}_{\text{mix}}^{t,r;\hat{t},0}}.$$

Proof. For $0 = T \leq \frac{r'-r}{t-t'}$ Lemma 2.7 yields the relation

$$M = |V_{L;\tilde{J}(L)}^{0;0}| \lesssim 2^{DL(1+c)}(cL^2)^{N-1},$$

where $c := (t+r)/\hat{t}$. With the assumptions of the lemma the relation (2.53) holds. Therefore, the estimate

$$\inf_{\tilde{f} \in V_{L;\tilde{J}(L)}^{0;0}} \|f - \tilde{f}\|_{\mathcal{H}_{\text{mix}}^{t',r'}} \lesssim 2^{L(t'+r'-(t+r))} \left(\|f\|_{\mathcal{H}_{\text{mix}}^{t,r}} + \|\hat{f}\|_{\mathcal{H}_{\text{mix}}^{\hat{t}}} \right)$$

$$\lesssim \left(\frac{2^{LD(1+c)} L^{2(N-1)}}{L^{2(N-1)}} \right)^{\frac{(t'+r'-(t+r))}{D(1+c)}} \left(\|f\|_{\mathcal{H}_{\text{mix}}^{t,r}} + \|\hat{f}\|_{\mathcal{H}_{\text{mix}}^{\hat{t}}} \right)$$

$$\lesssim \left(\frac{M}{\log_2(M)^{2(N-1)}} \right)^{\frac{(t'+r'-(t+r))}{D(1+c)}} \left(\|f\|_{\mathcal{H}_{\text{mix}}^{t,r}} + \|\hat{f}\|_{\mathcal{H}_{\text{mix}}^{\hat{t}}} \right)$$

follows. □

Thus, the convergence rate is up to logarithmic terms independet of the number particles. Note however, that due to possibly large terms $\|f\|_{\mathcal{H}_{\text{mix}}^{t,r}}$, $\|\hat{f}\|_{\mathcal{H}_{\text{mix}}^{\hat{t}}}$ and the constants involved in the approximation and complexity order estimates, the generalized sparse grid discretization scheme is only practical for a moderate number of low-dimensional particles. This holds even if there are no exponentially or logarithmically dependent terms with respect to the discretization parameter L present.

2.4 Weighted many-particle spaces

We discuss certain function spaces, i.e. so-called weighted spaces, which are related to a certain particle-wise decomposition of an N-particle function. Such a decomposition of an N-particle function $f : (\mathbb{R}^D)^N \to \mathbb{C}$ with respect to appropriate one-particle functions g_p, $p = 1, \ldots, N$ reads as

$$\begin{aligned} f(\vec{\mathbf{x}}) =& f_\emptyset \prod_{q \in \mathcal{N}} g_q(\mathbf{x}_q) \\ &+ \sum_{p_1 \in \mathcal{N}} f_{\{p_1\}}(\mathbf{x}_{p_1}) \prod_{q \in \mathcal{N}\setminus\{p_1\}} g_q(\mathbf{x}_q) \\ &+ \sum_{p_1 < p_2 \in \mathcal{N}} f_{\{p_1,p_2\}}(\mathbf{x}_{p_1}, \mathbf{x}_{p_2}) \prod_{q \in \mathcal{N}\setminus\{p_1,p_2\}} g_q(\mathbf{x}_q) \\ &+ \sum_{p_1 < p_2 < p_3 \in \mathcal{N}} f_{\{p_1,p_2,p_3\}}(\mathbf{x}_{p_1}, \mathbf{x}_{p_2}, \mathbf{x}_{p_3}) \prod_{q \in \mathcal{N}\setminus\{p_1,p_2,p_3\}} g_q(\mathbf{x}_q) \\ &\ldots \\ &+ f_\mathcal{N}(\mathbf{x}_1, \ldots, \mathbf{x}_N) \\ =& \sum_{u \subset \mathcal{N}} f_u(\vec{\mathbf{x}}_u) \prod_{p \in \mathcal{N}\setminus u} g_p(\mathbf{x}_p), \end{aligned} \quad (2.57)$$

where

$$\mathcal{N} := \{1, \ldots, N\}$$

and $\vec{x}_{\{p_1,\ldots,p_\mu\}}$ denotes the vector of variables $(\mathbf{x}_{p_1},\ldots,\mathbf{x}_{p_\mu})$ with $p_1 < \cdots < p_\mu$. In particular, a function f_u, $u \subset \mathcal{N}$ is a $|u|$-particle function and depends only on coordinates $\mathbf{x}_p$ with $p \in u$. Such a type of decomposition is known in the field of computational chemistry as a high-dimensional model representation (HDMR) [156] and is well-known in statistics under the name ANOVA (analysis of variance) [48]. In the following we discuss such types of decompositions. Note that the decomposition is exact and involves 2^N different terms due to the power set construction.

In general, it may happen that all terms in the decomposition are equally important or, alternatively, that the N-particle function $f_\mathcal{N}$ is the only important one. Then, nothing is gained from this decomposition. However, if, e.g. in the $\mathcal{H}_{\text{mix}}^{t,r}$-norm, the size of terms $\|f_u\|_{\mathcal{H}_{\text{mix}}^{t,r}}$ decays fast with e.g. $\mu = |u|$, then a proper restriction on certain low-particle functions of finite sum (2.57) results in a substantial reduction in computational complexity.

For a further reading on so-called ANOVA-like decompositions and weighted reproducing kernel Hilbert spaces see e.g. [64, 112, 156]. With respect to theoretical results on tractability of multivariate integration, linear multivariate problems and quasilinear multivariate problems in the framework of spaces with product or finite-order weights see e.g. [167, 184, 185, 188].

2.4.1 Particle-wise subspace splitting

A generalization of a decomposition as in (2.57) can be gained by a tensor product construction and a splitting of the one-particle space.

To this end, let $V^{(D)}$ denote a separable Hilbert space of D-dimensional functions and let $U_1,\ldots,U_N \subset V^{(D)}$. Let further W_p be function spaces such that we have a direct sum decomposition $V^{(D)} = U_p \oplus W_p$, for all $p = 1,\ldots,N$. Here, the splittings of the separable Hilbert space $V^{(D)}$ introduce a natural decomposition of the tensor product space $V^{(D,N)} := \bigotimes_{p=1}^N V^{(D)}$ by

$$V^{(D,N)} = \bigotimes_{p=1}^N (U_p \oplus W_p) = \bigoplus_{u \subset \mathcal{N}} \mathcal{W}_u, \qquad (2.58)$$

where

$$\mathcal{W}_u := \bigotimes_{p=1}^N \mathcal{W}_{u,(p)}, \quad \mathcal{W}_{u,(p)} := \begin{cases} U_p & \text{for } p \in \mathcal{N} \setminus u, \\ W_p & \text{for } p \in u. \end{cases} \qquad (2.59)$$

Moreover, for $p = 1,\ldots,N$, let $P_p : V^{(D)} \to U_p$ and let $Q_p : V^{(D)} \to W_p$ be linear projections, i.e. idempotent linear transformations, such that the identity operator $I^{(D,N)} : V^{(D,N)} \to V^{(D,N)}$ separates to $\bigotimes_{p=1}^N (P_p + Q_p)$. This leads to the splitting

$$I^{(D,N)} = \bigotimes_{p=1}^N (P_p + Q_p) = \sum_{u \subset \mathcal{N}} \mathcal{P}_u, \qquad (2.60)$$

where

$$\mathcal{P}_u := \bigotimes_{p=1}^N \mathcal{P}_{u,(p)}, \quad \mathcal{P}_{u,(p)} := \begin{cases} P_p & \text{for } p \in \mathcal{N} \setminus u, \\ Q_p & \text{for } p \in u. \end{cases} \qquad (2.61)$$

Accordingly, a function $f \in V^{(D,N)}$ is uniquely decomposed as

$$f(\vec{\mathbf{x}}) = \sum_{u \subset \mathcal{N}} F_u(\vec{\mathbf{x}}) \quad \text{with } F_u(\vec{\mathbf{x}}) := \mathcal{P}_u[f](\vec{\mathbf{x}}), \tag{2.62}$$

where $F_u \in \mathcal{W}_u$.

Now, let us assume that the subspaces $\{W_p\}_{p \in \mathcal{N}}$ and linear projections $\{Q_p\}_{p \in \mathcal{N}}$ are given by

$$W_p := \{Q_p[f] : f \in V^{(D)}\}, \quad Q_p := I - P_p, \tag{2.63}$$

where I denotes the identity operator on $V^{(D)}$. Then, the identity operator $I^{(D,N)}$ particularly splits according to (2.60). Let us further introduce the linear projections

$$P_u[f] := \bigotimes_{p=1}^{N} P_{u,(p)}[f], \quad P_{u,(p)} := \begin{cases} P_p & \text{for } p \in \mathcal{N} \setminus u, \\ I & \text{for } p \in u. \end{cases} \tag{2.64}$$

Now, we can obtain the functions in (2.62) recursively as

$$\begin{aligned} F_\emptyset &= P_\emptyset[f], \\ F_u &= P_u[f] - \sum_{\mu=0}^{|u|-1} \sum_{v \subset u, |v| = \mu} F_v. \end{aligned} \tag{2.65}$$

Note that the explicit formula

$$F_u = \sum_{v \subset u} (-1)^{|u|-|v|} P_v[f]$$

can also be deduced; see [112].

First order splitting

To gain a decomposition as in (2.57), we here consider the case that each subspace U_p is given by the span of a normalized one-particle function, i.e. $U_p := \text{span}\{g_p\}$, $g_p \in V^{(D)} \setminus \mathbf{0}$, $\|g_p\|_{V^{(D)}} = 1$, for $p = 1, \ldots, N$. Let further the linear projections $\{P_p\}_{p \in \mathcal{N}}$ be given by

$$P_p : V^{(D)} \to U_p : v \mapsto \langle g_p, v \rangle_{V^{(D)}} g_p \tag{2.66}$$

and let the subspaces $\{W_p\}_{p \in \mathcal{N}}$ and linear projections $\{Q_p\}_{p \in \mathcal{N}}$ be given analogously to (2.63). Then, W_p is orthogonal to U_p and we have an orthogonal direct sum decomposition $V^{(D)} = U_p \oplus W_p$, for all $p = 1, \ldots, N$. In particular, we may write the linear projections P_u, given in (2.64), in the form

$$P_u[f](\vec{\mathbf{x}}) = \underbrace{\left\langle \bigotimes_{p \in \mathcal{N} \setminus u} g_p, f \right\rangle_{\bigotimes_{p \in \mathcal{N} \setminus u} V^{(D)}} (\vec{\mathbf{x}}_u)}_{=: \tilde{P}_u[f](\vec{\mathbf{x}}_u)} \bigotimes_{p \in \mathcal{N} \setminus u} g_p(\vec{\mathbf{x}}_{\mathcal{N} \setminus u}). \tag{2.67}$$

Then, corresponding to (2.62) and (2.67), a function $f \in V^{(D,N)}$ is decomposed as

$$f(\vec{\mathbf{x}}) = \sum_{u \subset \mathcal{N}} F_u(\vec{\mathbf{x}}) = \sum_{u \subset \mathcal{N}} f_u(\vec{\mathbf{x}}_u) \prod_{p \in \mathcal{N} \setminus u} g_p(\mathbf{x}_p), \quad f_u(\vec{\mathbf{x}}_u) := \tilde{P}_u[F_u](\vec{\mathbf{x}}_u), \tag{2.68}$$

where $F_u \in \mathcal{W}_u$, $f_u \in W_u := \bigotimes_{p \in u} W_p$ and we particularly have the orthogonality relation $\langle F_u, F_v \rangle_{V^{(D,N)}} = 0$ for $u \neq v$. Here, decomposition (2.68) is unique for a fixed choice of $V^{(D)}$-normalized functions $g_1, \ldots, g_N \in V^{(D)}$, where we may obtain the functions F_u in (2.68) by (2.65). Note also that the functions f_u may be obtained by an application of the linear transformations

$$\tilde{\mathcal{P}}_u : V^{(D,N)} \to W_u$$

given for $u \subset \mathcal{N}$ by

$$\tilde{\mathcal{P}}_u[f](\vec{\mathbf{x}}_u) := \tilde{\mathcal{P}}_u[F_u](\vec{\mathbf{x}}_u) = f_u(\vec{\mathbf{x}}_u). \tag{2.69}$$

Let us remark that in the case of the space of square integrable functions on the one-dimensional torus $\mathbb{T}_1 = [-\frac{1}{2}, \frac{1}{2}]$, i.e. $V^{(1)} = \mathcal{L}^2(\mathbb{T}_1)$, and constant functions $g_1 = \cdots = g_N = 1$, the subspace splitting (2.58) and the decomposition (2.68) correspond with (2.66) and (2.63) just to the ANOVA decomposition; see e.g. [64].

Furthermore, let us shortly discuss the particle-wise decomposition of the N-particle space $\mathcal{L}^2((\mathbb{R}^D)^N)$ with respect to arbitrarily given $\mathcal{L}^2$-normalized one-particle functions $g_1, \ldots, g_N \in \mathcal{L}^2(\mathbb{R}^D)$. Here, we obtain according to (2.64) and (2.66) the linear projections P_u in the form

$$P_u[f](\vec{\mathbf{x}}) = \underbrace{\left(\int_{(\mathbb{R}^D)^{N-|u|}} \left(\prod_{p \in \mathcal{N} \setminus u} g_p^*(\mathbf{y}_p) \right) f(\vec{\mathbf{y}})|_{\vec{\mathbf{y}} = \vec{\mathbf{x}} \setminus \vec{\mathbf{y}}_{\mathcal{N} \setminus u}} \, d\vec{\mathbf{y}}_{\mathcal{N} \setminus u} \right)}_{=\tilde{P}_u[f](\vec{\mathbf{x}}_u)} \left(\prod_{p \in \mathcal{N} \setminus u} g_p(\mathbf{x}_p) \right), \tag{2.70}$$

where we use the notation

$$f(\vec{\mathbf{y}})|_{\vec{\mathbf{y}} = \vec{\mathbf{x}} \setminus \vec{\mathbf{y}}_{\{p\}}} = f(\mathbf{x}_1, \ldots, \mathbf{x}_{p-1}, \mathbf{y}_p, \mathbf{x}_{p+1}, \ldots, \mathbf{x}_N)$$

and its obvious generalization to $\vec{\mathbf{x}} \setminus \vec{\mathbf{y}}_u$. Then the functions F_u of the particle-wise decomposition can be written according to (2.65), (2.68) and (2.70) in the form

$$F_\emptyset(\vec{\mathbf{x}}) = \int_{(\mathbb{R}^D)^N} \left(\prod_{p \in \mathcal{N}} g_p^*(\mathbf{y}_p) \right) f(\vec{\mathbf{y}}) \, d\vec{\mathbf{y}} \prod_{p \in \mathcal{N}} g_p(\mathbf{x}_p),$$

$$F_{\{p_1\}}(\vec{\mathbf{x}}) = \int_{(\mathbb{R}^D)^{N-1}} \left(\prod_{p \in \mathcal{N} \setminus \{p_1\}} g_p^*(\mathbf{y}_p) \right) f(\vec{\mathbf{y}})|_{\vec{\mathbf{y}} = \vec{\mathbf{x}} \setminus \vec{\mathbf{y}}_{\mathcal{N} \setminus \{p_1\}}} \, d\vec{\mathbf{y}}_{\mathcal{N} \setminus \{p_1\}} \prod_{p \in \mathcal{N} \setminus \{p_1\}} g_p(\mathbf{x}_p)$$
$$- F_\emptyset(\vec{\mathbf{x}}),$$

$$F_{\{p_1, p_2\}}(\vec{\mathbf{x}}) = \tilde{P}_{\{p_1, p_2\}}[f](\vec{\mathbf{x}}_{\{p_1, p_2\}}) \prod_{p \in \mathcal{N} \setminus \{p_1, p_2\}} g_p(\mathbf{x}_p)$$
$$- F_{\{p_1\}}(\vec{\mathbf{x}}) - F_{\{p_2\}}(\vec{\mathbf{x}})$$
$$- F_\emptyset(\vec{\mathbf{x}}),$$
$$\cdots \cdots$$

$$F_u(\vec{\mathbf{x}}) = \tilde{P}_u[f](\vec{\mathbf{x}}_u) \prod_{p \in \mathcal{N} \setminus u} g_p(\mathbf{x}_p) - \sum_{\mu=0}^{|u|-1} \sum_{v \subset u, |v| = \mu} F_v(\vec{\mathbf{x}}),$$
$$\cdots \cdots$$

So far we have not discussed how to choose the normalized one-particle functions $g_1, \ldots, g_N$. Here, in order to gain a fast decay in the size of the terms $\|f_u\|_{\mathcal{L}^2}$ for

2 Sobolev Spaces for Many-Particle Functions

example the approximation error of $\|f - F_\emptyset\|_{\mathcal{L}^2}$ might be minimized. Thereby, we employ normalized one-particle functions which correspond to a best rank-1 approximation with respect to the $\mathcal{L}^2$-norm

$$\|f - F_\emptyset\|_{\mathcal{L}^2} = \|f - f_\emptyset \bigotimes_{p=1}^{N} g_p\|_{\mathcal{L}^2}$$

$$= \min_{\tilde{g}_1, \ldots, \tilde{g}_N} \|f - \bigotimes_{p=1}^{N} \tilde{g}_p\|_{\mathcal{L}^2}.$$

Higher order splittings

Let us now consider subspace splittings as in (2.58), where each of the subspaces U_p is given by the span of a finite set of linear independent one-particle functions.

Here, we restrict ourselves on the case that all subspaces $\{U_p\}_{p \in \mathcal{N}}$ and all subspaces $\{W_p\}_{p \in \mathcal{N}}$ are equal to a subspace $U = U_1 = \cdots = U_N$ and a subspace $W = W_1 = \cdots = W_N$, respectively. Then, we particularly have the direct sum decomposition $V^{(D)} = U \oplus W$ and the subspace splitting of the tensor product space $V^{(D,N)} = \bigotimes_{p=1}^{N}(U \oplus W) = \bigoplus_{u \subset \mathcal{N}} \mathcal{W}_u$, with $\mathcal{W}_u$ from (2.59). Now, let U be spanned by a given set $\{g_1, \ldots, g_{N_U}\}$ of linear independent functions and let us introduce for pairwise disjoint $u_1, \ldots, u_{N_U} \subset \mathcal{N}$ the spaces

$$\mathcal{U}_{u_1,\ldots,u_{N_U}} := \bigotimes_{p=1}^{N} \mathcal{U}_{u_1,\ldots,u_{N_U},(p)}, \quad \mathcal{U}_{u_1,\ldots,u_{N_U},(p)} := \begin{cases} \text{span}\{g_1\} & \text{for } p \in u_1, \\ \vdots \\ \text{span}\{g_{N_U}\} & \text{for } p \in u_{N_U}, \\ W & \text{for } p \in \mathcal{N} \setminus \bigcup_{\nu=1}^{N_U} u_\nu. \end{cases} \quad (2.71)$$

Then, a natural subspace splitting is given by[6]

$$V^{(D,N)} = \bigotimes_{p=1}^{N}(U \oplus W) = \bigoplus_{u_1,\ldots,u_{N_U} \subset \mathcal{N}, \forall \nu \neq \nu': u_\nu \cap u_{\nu'} = \emptyset} \mathcal{U}_{u_1,\ldots,u_{N_U},(p)}. \quad (2.72)$$

In the same way as in the first order case we can separate the identity operator with appropriate linear projections to obtain a decomposition of an N-particle function in the form

$$f(\vec{\mathbf{x}}) = \sum_{u_1,\ldots,u_{N_U} \subset \mathcal{N}, \forall \nu \neq \nu': u_\nu \cap u_{\nu'} = \emptyset} \tilde{F}_{u_1,\ldots,u_{N_U}}(\vec{\mathbf{x}})$$

$$= \sum_{u_1,\ldots,u_{N_U} \subset \mathcal{N}, \forall \nu \neq \nu': u_\nu \cap u_{\nu'} = \emptyset} \tilde{f}_{u_1,\ldots,u_{N_U}}(\vec{\mathbf{x}}_{\mathcal{N} \setminus \bigcup_{\nu=1}^{N_U} u_\nu}) \prod_{\nu=1}^{N_U} \prod_{p_\nu \in u_\nu} g_{p_\nu}(\mathbf{x}_{p_\nu}), \quad (2.73)$$

---

[6]Note that for finite index sets u, v the identity

$$\prod_{p \in u} \left(\sum_{q \in v} c_{p,q} \right) = \sum_{\{u_q\}_{q \in v}, \forall q \neq q': u_q \cap u_{q'} = \emptyset, \bigcup_{q \in v} u_q = u} \prod_{q \in v} \prod_{p \in u_q} c_{p,q}$$

holds, where $c_{p,q} \in \mathbb{C}$. This can be generalized in a straightforward way to products of sums of functions and products of sums of subspaces.

with $\tilde{F}_{u_1,\ldots,u_{N_U}} \in \mathcal{U}_{u_1,\ldots,u_{N_U}}$ and $\tilde{f} \in \bigotimes_{p \in \mathcal{N} \setminus \bigcup_{\nu=1}^{N_U} u_\nu}$.[7] We discuss this in more detail for the specific case of antisymmetric many-particle spaces in Section 2.5.2.

Moreover, we have not discussed how to choose the linear independent one-particle functions $g_1, \ldots, g_N$. Obviously, one possibility is to take them from an orthonormal basis set $\mathcal{B}_{V^{(D)}} = \{e_\nu\}_{\nu \in \Lambda \subset \mathbb{Z}}$ of the separable Hilbert space $V^{(D)}$, i.e. $g_1, \ldots, g_N \in \mathcal{B}_{V^{(D)}}$. However, the choice of the one-particle functions from the different orthonormal basis sets of the separable Hilbert space $V^{(D)}$ is not obvious. Probably, this choice is problem-specific. In the case of antisymmetric N-particle functions we refer to the discussion in Section 2.5.2 and Section 4.3.1.

2.4.2 Many-particle spaces with finite-order weights

Let us first discuss the case of a first order splitting of the tensor product space $V^{(D,N)} = \bigotimes_{p=1}^{N} V^{(D)}$, where $V^{(D)}$ denotes a separable Hilbert space. To this end, let $g_1, \ldots, g_N \in V^{(D)}$ be $V^{(D)}$-normalized one-particle functions and let further the linear projections $\{P_p\}_{p \in \mathcal{N}}$, $\{Q_p\}_{p \in \mathcal{N}}$ and subspaces $\{W_p\}_{p \in \mathcal{N}}$ be given by (2.66) and (2.63). In this way we obtain a subspace splitting of the form (2.58) with $U_p = \text{span}\{g_p\}$, and we can uniquely decompose any N-particle function $f \in V^{(D,N)}$ according to (2.68) with the help of the linear transformations $\tilde{\mathcal{P}}_u$ in (2.69). Since the one-particle functions $\{g_p\}_{p \in \mathcal{N}}$ are normalized and since the orthogonality relation holds, i.e. $\langle F_u, F_v \rangle_{V^{(D,N)}} = 0$ for $u \neq v$, the $V^{(D,N)}$-norm splits similar to decomposition (2.68)

$$\|f\|^2_{V^{(D,N)}} = \sum_{u \subset \mathcal{N}} \|F_u\|^2_{V^{(D,N)}} = \sum_{u \subset \mathcal{N}} \|f_u\|^2_{\bigotimes_{p \in u} V^{(D)}}. \quad (2.74)$$

Now, let $h \in V^{(D,N)}$ be given by $h(\vec{x}) = \sum_{u \subset \mathcal{N}} h_u(\vec{x}_u) \prod_{p \in \mathcal{N} \setminus u} g_p(\mathbf{x}_p)$ with $h_u \in \bigotimes_{p \in u} W_p$. Then the error, measured in the $V^{(D,N)}$-norm, of approximating a function $f \in V^{(D,N)}$ by the function h is related to the sum of the errors, measured in the $\left(\bigotimes_{p \in u} V^{(D)}\right)$-norm, of approximating the functions $\tilde{\mathcal{P}}_u[f]$ by h_u, i.e.

$$\|f - h\|^2_{V^{(D,N)}} = \sum_{u \subset \mathcal{N}} \|\tilde{\mathcal{P}}_u[f] - h_u\|^2_{\bigotimes_{p \in u} V^{(D)}}.$$

Let us remark that according to (2.14) we obtain the Sobolev space of dominant mixed smoothness $V^{(D,N)} = \mathcal{H}^t_{\text{mix}}(\mathbb{R}^D)$ with $t \geq 0$, if we set $V^{(D)} = \mathcal{H}^t(\mathbb{R}^D)$. In this way, a first order subspace splitting of $\mathcal{H}^t_{\text{mix}}((\mathbb{R}^D)^N)$ is given by (2.58) and correspondingly a decomposition of $f \in \mathcal{H}^t_{\text{mix}}$ and of its norm by (2.68) and (2.74).

In the following we focus on subspaces of the N-particle space $\mathcal{L}^2((\mathbb{R}^D)^N)$ only, i.e. we set $V^{(D)} = \mathcal{L}^2(\mathbb{R}^D)$ and hence $V^{(D,N)} = \mathcal{L}^2((\mathbb{R}^D)^N)$. For $t + r \geq 0$, $t \geq 0$ and non-negative weights $\gamma_u \geq 0$, for all $u \subset \mathcal{N}$, we introduce within this framework

$$\|f\|^2_{\mathcal{H}^{t,r}_{\text{mix},\{g_p\}_{p \in \mathcal{N}},\{\gamma_u\}_{u \subset \mathcal{N}}}} := \sum_{u \subset \mathcal{N}} \gamma_u^{-1} \|\tilde{\mathcal{P}}_u[f]\|^2_{\mathcal{H}^{t,r}_{\text{mix}}} = \sum_{u \subset \mathcal{N}} \gamma_u^{-1} \|f_u\|^2_{\mathcal{H}^{t,r}_{\text{mix}}} \quad (2.75)$$

with

$$\mathcal{H}^{t,r}_{\text{mix},\{g_p\}_{p \in \mathcal{N}},\{\gamma_u\}_{u \subset \mathcal{N}}}((\mathbb{R}^D)^N) := \left\{ f \in \mathcal{L}^2((\mathbb{R}^D)^N) : \|f\|^2_{\mathcal{H}^{t,r}_{\text{mix},\{g_p\},\{\gamma_u\}}} < \infty \right\}. \quad (2.76)$$

---

[7] Note that in the case of $N_U = 1$ the decomposition (2.73) corresponds to a first order decomposition according to (2.57), where $F_u = \tilde{F}_{\mathcal{N} \setminus u}$, $f_u = \tilde{f}_{\mathcal{N} \setminus u}$ and $g_1 = \cdots = g_N$.

With respect to the approximation of a function in $\mathcal{H}^{t,r}_{\mathrm{mix},\{g_p\}_{p\in\mathcal{N}},\{\gamma_u\}_{u\subset\mathcal{N}}}$, we give an estimate in the following lemma:

Lemma 2.9. *Let $t + r \geq 0$, $t \geq 0$, $\{\gamma_u\}_{u\subset\mathcal{N}}$ be a set of non-negative weights, $f \in \mathcal{H}^{t,r}_{\mathrm{mix},\{g_p\}_{p\in\mathcal{N}},\{\gamma_u\}_{u\subset\mathcal{N}}}((\mathbb{R}^D)^N)$ and let the one-particle functions $g_1, \ldots, g_N \in \mathcal{H}^{t+r}(\mathbb{R}^D)$ be $\mathcal{L}^2$-normalized. Let further*

$$h(\vec{\mathbf{x}}) = \sum_{u\subset\mathcal{N}} h_u(\vec{\mathbf{x}}_u) \prod_{p\in\mathcal{N}\setminus u} g_p(\mathbf{x}_p) \quad \text{with } h_u \in \mathcal{H}^{t,r}_{\mathrm{mix}}((\mathbb{R}^D)^{|u|}) \cap \bigotimes_{p\in u} W_p.$$

Then the estimate

$$\|f - h\|_{\mathcal{H}^{t,r}_{\mathrm{mix}}} \lesssim \sum_{u\subset\mathcal{N}} \|f_u - h_u\|^2_{\mathcal{H}^{t,r}_{\mathrm{mix}}}$$

holds, where $f_u = \tilde{\mathcal{P}}_u[f]$.

Proof. Since $\|f\|^2_{\mathcal{H}^{t,r}_{\mathrm{mix},\{g_p\},\{\gamma_u\}}} < \infty$ it follows for all $u \subset \mathcal{N}$ that $f_u \in \mathcal{H}^{t,r}_{\mathrm{mix}}$. Therefore, with $g_p \in \mathcal{H}^{t+r}(\mathbb{R}^D)$, we obtain $\mathcal{P}_u[f] = f_u \prod_{p\in\mathcal{N}\setminus u} g_p \in \mathcal{H}^{t,r}_{\mathrm{mix}}$, $h_u \prod_{p\in\mathcal{N}\setminus u} g_p \in \mathcal{H}^{t,r}_{\mathrm{mix}}$ and thus $f, h \in \mathcal{H}^{t,r}_{\mathrm{mix}}$. Then with the help of Definition 2.2 it easily follows that

$$\|f - h\|^2_{\mathcal{H}^{t,r}_{\mathrm{mix}}} \leq \sum_{u\subset\mathcal{N}} \|\mathcal{P}_u[f] - h_u \bigotimes_{p\in\mathcal{N}\setminus u} g_p\|^2_{\mathcal{H}^{t,r}_{\mathrm{mix}}}$$

$$\lesssim \sum_{u\subset\mathcal{N}} \|\tilde{\mathcal{P}}_u[f] - h_u\|^2_{\mathcal{H}^{t,r}_{\mathrm{mix}}((\mathbb{R}^D)^{|u|})} \|\bigotimes_{p\in\mathcal{N}\setminus u} g_p\|^2_{\mathcal{H}^{t,r}_{\mathrm{mix}}((\mathbb{R}^D)^{N-|u|})}$$

and hence we obtain the desired result. $\square$

Note that for $f \in \mathcal{H}^{t,r}_{\mathrm{mix},\{g_p\}_{p\in\mathcal{N}},\{\gamma_u\}_{u\subset\mathcal{N}}}$ the weight γ_u prescribes the importance of the term f_u and hence the importance of different particles and of correlations between groups of particles. In particular for a weight $\gamma_u \to 0$ the norm $\|f_u\|_{\mathcal{H}^{t,r}_{\mathrm{mix}}}$ is forced to be zero. Therefore, by a weight equal to zero, $\gamma_u = 0$, the associated subspace $\mathcal{W}_u$ is switched off in the first order subspace splitting. Let us now consider $\mathcal{H}^{t,r}_{\mathrm{mix},\{g_p\}_{p\in\mathcal{N}},\{\gamma_u\}_{u\subset\mathcal{N}}}((\mathbb{R}^D)^N)$ equipped with finite-order weights. A set of weights $\{\gamma_u\}_{u\subset\mathcal{N}}$ is a so-called set of finite-order weights of order q if $\gamma_u = 0$ for all γ_u with $|u| > q$. Then the inclusion $\mathcal{H}^{t,r}_{\mathrm{mix},\{g_p\}_{p\in\mathcal{N}},\{\gamma_u\}_{u\subset\mathcal{N}}} \subset \bigoplus_{\mu=0}^{q} \bigoplus_{u\subset\mathcal{N},|u|=\mu} \mathcal{W}_u$ holds and the problem of the approximation of an N-particle function $f \in \mathcal{H}^{t,r}_{\mathrm{mix},\{g_p\}_{p\in\mathcal{N}},\{\gamma_u\}_{u\subset\mathcal{N}}}$ can be reduced to the problem of approximating 0-particle, $\binom{N}{1}$ 1-particle, ..., $\binom{N}{q}$ q-particle functions in $\mathcal{H}^{t,r}_{\mathrm{mix}}$. In particular, with the assumptions of Lemma 2.9, we obtain

$$\|f - h\|_{\mathcal{H}^{t,r}_{\mathrm{mix}}} \lesssim \sum_{\mu=0}^{q} \binom{N}{\mu} \max_{u\subset\mathcal{N},|u|=\mu} \|f_u - h_u\|^2_{\mathcal{H}^{t,r}_{\mathrm{mix}}}.$$

If we additionally assume that $f_u \in \mathcal{H}^{t,r;\hat{t},\hat{r}}_{\mathrm{mix}}$ for all $u \subset \mathcal{N}$, we can apply a generalized sparse grid discretization scheme for the functions f_u.[8] Then, the order terms and

---

[8] Note that we may adapt (2.75) and (2.76) to the case of $\mathcal{H}^{t,r;\hat{t},\hat{r}}_{\mathrm{mix}}$ for $\hat{t} + \hat{r} \geq 0$, $\hat{t} \geq 0$ by

$$\|f\|^2_{\mathcal{H}^{t,r;\hat{t},\hat{r}}_{\mathrm{mix},\{g_p\}_{p\in\mathcal{N}},\{\gamma_u\}_{u\subset\mathcal{N}}}} := \sum_{u\subset\mathcal{N}} \gamma_u^{-1} \|\tilde{\mathcal{P}}_u[f]\|^2_{\mathcal{H}^{t,r;\hat{t},\hat{r}}_{\mathrm{mix}}} = \sum_{u\subset\mathcal{N}} \gamma_u^{-1} \left(\|f_u\|^2_{\mathcal{H}^{t,r}_{\mathrm{mix}}} + \|\mathcal{F}[f_u]\|^2_{\mathcal{H}^{\hat{t},\hat{r}}_{\mathrm{mix}}} \right)$$

the constants in the approximation and complexity order estimates with respect to the discretization parameter L, i.e. Lemma 2.4, Lemma 2.5 and 2.7, dependent on q instead of N. Therefore, if the order q is sufficiently small, then the generalized sparse grid discretization scheme is practically independent of N.

Let us remark that if a subspace splitting $\bigotimes_{p=1}^{N}(U_p + W_p) = \sum_{u \subset \mathcal{N}} \mathcal{W}_u$, i.e. a non unique decomposition, is considered, we can use

$$\|f\|^2_{\mathcal{H}^{t,r;\inf}_{\mathrm{mix},\{g_p\}_{p\in\mathcal{N}},\{\gamma_u\}_{u\subset\mathcal{N}}}} := \inf_{\{f_u \in \mathcal{W}_u\}_{u\subset\mathcal{N}}, f = \sum_{u\subset\mathcal{N}} f_u \prod_{p\in\mathcal{N}\setminus u} g_p} \sum_{u\subset\mathcal{N}} \gamma_u^{-1} \|f_u\|^2_{\mathcal{H}^{t,r}_{\mathrm{mix}}}$$

instead of (2.75) to define weighted spaces. Note further, that spaces of finite-order weights can also be adapted to the case of higher order splittings by using a set of non-negative weights $\{\gamma_{u_1,\ldots,u_{N_U}}\}_{u_1,\ldots,u_{N_U} \subset \mathcal{N}, \forall \nu \neq \nu': u_\nu \cap u_{\nu'} = \emptyset}$; compare (2.72) and (2.73). Note finally, that besides finite-order weights also so-called product weights or more general weights may lead to tractability in the framework of weighted reproducing kernel Hilbert spaces; see e.g. [43, 64].

2.5 Symmetric and antisymmetric many-particle spaces

In quantum mechanics it is common to consider identical particles, i.e. particles which are indistinguishable from each other. Usually, for the quantum mechanical description of identical particles many-particles state functions are employed which obey certain symmetry conditions. Here, the choice of symmetry or antisymmetry is determined by the species of particle. Therefore, we shortly discuss Hilbert spaces for symmetric and antisymmetric many-particle functions.

Let V denote a separable Hilbert space of N-particle functions $f : \Omega \subset (\mathbb{R}^D)^N \to \mathbb{C}$, where we assume that Ω is invariant under all permutations, i.e.

$$\vec{x} \in \Omega \Rightarrow P\vec{x} \in \Omega, \forall P \in \mathcal{S}_N, \qquad (2.77)$$

and that for each N-particle function in V also all its permutations with respect to particle coordinates are in V, i.e.

$$f \in V \Rightarrow (\vec{x} \mapsto f(P\vec{x})) \in V, \forall P \in \mathcal{S}_N. \qquad (2.78)$$

Here, $\mathcal{S}_N$ denotes the symmetric group of degree N. In particular, a permutation P is a mapping $P : \{1,\ldots,N\} \to \{1,\ldots,N\}$ which translates to a permutation of the corresponding numbering of particles and thus to a permutation of indices, i.e. we have $P(\mathbf{x}_1,\ldots,\mathbf{x}_N) := (\mathbf{x}_{P(1)},\ldots,\mathbf{x}_{P(N)})$. Now, the subspace of antisymmetric functions in V is defined by

$$V_{\mathfrak{A}} := \left\{ f \in V : (-1)^{|P|} f(P\vec{x}) - f(\vec{x}) = 0, \forall P \in \mathcal{S}_N \right\} \subset V$$

and the subspace of symmetric functions in V is defined by

$$V_{\mathfrak{S}} := \{ f \in V : f(P\vec{x}) - f(\vec{x}) = 0, \forall P \in \mathcal{S}_N \} \subset V.$$

and

$$\mathcal{H}^{t,r;\hat{t},\hat{r}}_{\mathrm{mix},\{g_p\}_{p\in\mathcal{N}},\{\gamma_u\}_{u\subset\mathcal{N}}}((\mathbb{R}^D)^N) := \left\{ f \in \mathcal{L}^2((\mathbb{R}^D)^N) : \|f\|^2_{\mathcal{H}^{t,r;\hat{t},\hat{r}}_{\mathrm{mix},\{g_p\},\{\gamma_u\}}} < \infty \right\},$$

respectively.

In particular $V_\mathfrak{S} = \mathfrak{S}(V)$ and $V_\mathfrak{A} = \mathfrak{A}(V)$, where the so-called antisymmetrizer, i.e. a linear antisymmetrization operator, is given by

$$\mathfrak{A}: V \to V_\mathfrak{A}: f \mapsto \frac{1}{|\mathcal{S}_N|} \sum_{P \in \mathcal{S}_N} (-1)^{|P|} f(P\vec{x}) \tag{2.79}$$

and the so-called symmetrizer, i.e. a linear symmetrization operator, is given by

$$\mathfrak{S}: V \to V_\mathfrak{S}: f \mapsto \frac{1}{|\mathcal{S}_N|} \sum_{P \in \mathcal{S}_N} f(P\vec{x}).$$

Here, $|P|$ denotes the inversion number of the permutation P and thus the expression $(-1)^{|P|}$ is equal to the sign, or parity, of the permutation $\text{sgn}(P)$ and also equal to the determinant of the associated permutation matrix $|P|$. Let us remark that the antisymmetrizer and the symmetrizer are linear projections, since they are linear transformations and idempotent, i.e. $\mathfrak{A}^2 = \mathfrak{A}$ and $\mathfrak{S}^2 = \mathfrak{S}$. Note further that, if the set $\{\phi_\mu\}_{\mu \in \Lambda \subset \mathbb{Z}}$ generates the space V, then the subspaces $V_\mathfrak{A}$ and $V_\mathfrak{S}$ are spanned by the sets $\{\mathfrak{A}\phi_\mu\}_{\mu \in \Lambda}$ and $\{\mathfrak{S}\phi_\mu\}_{\mu \in \Lambda}$, respectively.[9]

Let us discuss the case that V is a subspace of the tensor product space $V^{(D,N)} = \bigotimes_{p=1}^N V^{(D)}$, where $V^{(D)}$ is a separable Hilbert space of D-dimensional functions. To this end, let $\mathcal{B}_{V^{(D)}} = \{\mathbf{b}_\mu\}_{\mu \in \Lambda_{(D)}}$ be a basis of $V^{(D)}$, where $\Lambda_{(D)} \subset \mathbb{Z}$. Hence, the set $\mathcal{B}_{V^{(D,N)}} = \{\vec{\mathbf{b}}_{\vec{\mu}}\}_{\vec{\mu} \in (\Lambda_{(D)})^N}$ forms a basis of $V^{(D,N)}$, where the N-particle functions $\vec{\mathbf{b}}_{\vec{\mu}}$ are given by the tensor product of the one-particle basis functions

$$\vec{\mathbf{b}}_{\vec{\mu}} := \bigotimes_{p=1}^N \mathbf{b}_{\mu_p}.$$

Now, we assume that the subset of the N-particle functions $\mathcal{B}_V = \{\vec{\mathbf{b}}_{\vec{\mu}}\}_{\vec{\mu} \in \Lambda \subset (\Lambda_{(D)})^N} \subset \mathcal{B}_{V^{(D,N)}}$ builds a basis for V, where we assume an analogous property as in (2.78) for the index set Λ, i.e.

$$\vec{\mu} \in \Lambda \Rightarrow P\vec{\mu} \in \Lambda, \forall P \in \mathcal{S}_N. \tag{2.80}$$

Then, the sequences $\{\mathfrak{A}\vec{\mathbf{b}}_{\vec{\mu}}\}_{\vec{\mu} \in \Lambda}$ and $\{\mathfrak{S}\vec{\mathbf{b}}_{\vec{\mu}}\}_{\vec{\mu} \in \Lambda}$ form generating systems of the antisymmetric subspace $V_\mathfrak{A} \subset V$ and the symmetric subspace $V_\mathfrak{S} \subset V$, respectively. However, many functions in the sequence $\{\mathfrak{S}\vec{\mathbf{b}}_{\vec{\mu}}\}_{\vec{\mu} \in \Lambda}$ are identical and many functions in the sequence $\{\mathfrak{A}\vec{\mathbf{b}}_{\vec{\mu}}\}_{\vec{\mu} \in \Lambda}$ are identical up to sign.

We can gain a basis for the subspace $V_\mathfrak{S}$ if we restrict the sequence $\{\mathfrak{S}\vec{\mathbf{b}}_{\vec{\mu}}\}_{\vec{\mu} \in \Lambda}$ to one representative index of each equivalence class of Λ under the equivalence relation

$$\vec{\mu} \sim_\mathfrak{S} \vec{\mu}' :\Leftrightarrow \mathfrak{S}\vec{\mathbf{b}}_{\vec{\mu}} - \mathfrak{S}\vec{\mathbf{b}}_{\vec{\mu}'} = 0.$$

In the antisymmetric case, we can restrict the sequence $\{\mathfrak{A}\vec{\mathbf{b}}_{\vec{\mu}}\}_{\vec{\mu} \in \Lambda}$ in the same way using the equivalence relation

$$\vec{\mu} \sim_\mathfrak{A} \vec{\mu}' :\Leftrightarrow \mathfrak{A}\vec{\mathbf{b}}_{\vec{\mu}} - \mathfrak{A}\vec{\mathbf{b}}_{\vec{\mu}'} = 0 \text{ or } \mathfrak{A}\vec{\mathbf{b}}_{\vec{\mu}} + \mathfrak{A}\vec{\mathbf{b}}_{\vec{\mu}'} = 0,$$

where we additionally exclude the equivalence class associated with the N-particle function $0 \in V$. These restrictions can be done in many different ways. Possible basis sets

---

[9] We assume that except for completion with respect to the norm $\|.\|_V = \sqrt{\langle .,. \rangle_V}$, the span of the set $\{\phi_\mu\}_{\mu \in \Lambda}$ is is just the associated Hilbert space V. Thus, the subspaces $V_\mathfrak{A}$ and $V_\mathfrak{S}$ are generated, up to completion, by the sets $\{\mathfrak{A}\phi_\mu\}_{\mu \in \Lambda}$ and $\{\mathfrak{S}\phi_\mu\}_{\mu \in \Lambda}$, respectively.

2.5 Symmetric and antisymmetric many-particle spaces

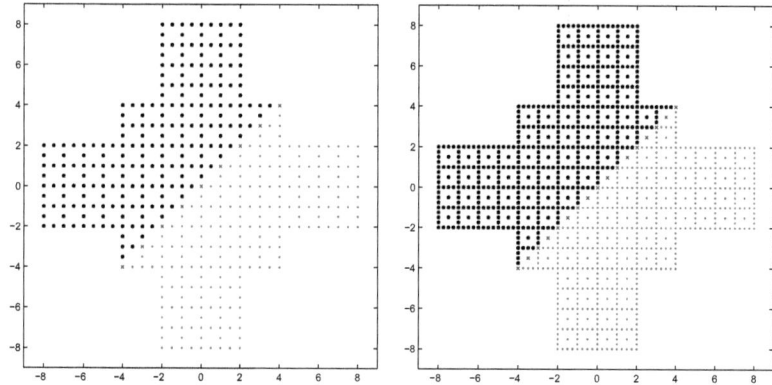

Figure 2.5: Localization peaks of basis functions in $V_{L;3}^{0;0}$ from (2.50) with $D = 1$, $N = 2$, $L = 2$ (left) and $L = 3$ (right). Here, the black dots represent indices of basis functions associated with the antisymmetric basis set $\mathcal{B}_{V_\mathfrak{A}}$ and the union of the black dots and the crosses on the diagonal represent indices of basis functions associated with the symmetric basis set $\mathcal{B}_{V_\mathfrak{S}}$.

for $V_\mathfrak{S}$ and $V_\mathfrak{A}$ are given by

$$\mathcal{B}_{V_\mathfrak{S}} := \left\{ \mathfrak{S}\vec{\mathbf{b}}_{\vec{\mu}} : \mu \in \Lambda, \mu_1 \leq \cdots \leq \mu_N \right\} \tag{2.81}$$

and

$$\mathcal{B}_{V_\mathfrak{A}} := \left\{ \mathfrak{A}\vec{\mathbf{b}}_{\vec{\mu}} : \mu \in \Lambda, \mu_1 < \cdots < \mu_N \right\}, \tag{2.82}$$

respectively. For examples of antisymmetric and symmetric general sparse grid spaces see Figure 2.5. Note that if the set $\mathcal{B}_{V^{(D)}} = \{\mathbf{b}_\mu\}_{\mu \in \Lambda_{(D)}}$ is a generating system only and not a basis set for $V^{(D)}$, then, analogously, the sets $\mathcal{B}_{V_\mathfrak{S}}$ and $\mathcal{B}_{V_\mathfrak{A}}$ generate the subspaces $V_\mathfrak{S}$ and $V_\mathfrak{A}$, respectively.

Let us now discuss the case that the one-particle space $V^{(D)}$ is spanned by a finite basis set $\mathcal{B}_{V^{(D)}} = \{\mathbf{b}_\mu\}_{\mu \in \Lambda_{(D)}}$ with $\Lambda_{(D)} \subset \mathbb{Z}$ and $|\Lambda_{(D)}| = m$. Then the cardinality numbers of the basis sets for the tensor product space $V^{(D,N)} = \bigotimes_{p=1}^{N} V^{(D)}$, the symmetric subspace $V_\mathfrak{S}^{(D,N)} \subset V^{(D,N)}$ and the antisymmetric subspace $V_\mathfrak{A}^{(D,N)} \subset V^{(D,N)}$ read as

$$\begin{aligned}
|\mathcal{B}_{V^{(D,N)}}| &= m^N, \\
|\mathcal{B}_{V_\mathfrak{S}^{(D,N)}}| &= \binom{m+N-1}{N} \leq \frac{(m+N-1)^N}{N!}, \\
|\mathcal{B}_{V_\mathfrak{A}^{(D,N)}}| &= \binom{m}{N} \leq \frac{m^N}{N!},
\end{aligned} \tag{2.83}$$

respectively.[10] In particular, if the subset $\mathcal{B}_V = \{\vec{\mathbf{b}}_{\vec{\mu}}\}_{\vec{\mu} \in \Lambda} \subset \mathcal{B}_{V^{(D,N)}}$ with property (2.80)

---

[10]Note that the binomial coefficient $\binom{n}{k}$ is bounded by $\left(\frac{n}{k}\right)^k \leq \binom{n}{k} \leq \frac{n^k}{k!}$.

and with $|\mathcal{B}_V| = M$ builds a basis for V, then the upper estimate

$$|\mathcal{B}_{V_\mathfrak{A}}| \leq \frac{M}{N!} \tag{2.84}$$

obviously holds for the cardinality of the basis set for the antisymmetric subspace $V_\mathfrak{A}$.

Let us further consider the case that $\mathcal{B}_{V^{(D)}} = \{\mathbf{b}_\mu\}_{\mu \in \Lambda_{(D)}}$ is an orthonormal basis set for $V^{(D)}$. Then the set

$$\left\{ \frac{1}{\sqrt{N! \prod_{p=1}^N N_p!}} \sum_{P \in \mathcal{S}_N} \vec{\mathbf{b}}_{P\vec{\mu}} : \mu \in \Lambda, \mu_1 \leq \cdots \leq \mu_N \right\}$$

is an orthonormal basis for the symmetric subspace $V_\mathfrak{S}$, where N_p denotes the number of times that each one-particle basis function is found in the N-particle basis function. In the following we discuss the antisymmetric case in more detail. Note that for any tensor product of one-particle functions $\phi_1, \ldots, \phi_N \in V^{(D)}$ we then have

$$\mathfrak{A} \bigotimes_{p=1}^N \phi_p(\vec{\mathbf{x}}) = \frac{1}{N!} \sum_{P \in \mathcal{S}_N} \bigotimes_{p=1}^N \phi_p(P\vec{\mathbf{x}}) = \frac{1}{N!} \sum_{P \in \mathcal{S}_N} \bigotimes_{p=1}^N \phi_{P(p)}(\vec{\mathbf{x}})$$

$$= \frac{1}{N!} \bigwedge_{p=1}^N \phi_p(\vec{\mathbf{x}}).$$

In other words, the classical tensor product of one-particle functions $\bigotimes_{p=1}^N \phi_p$ is up to a constant replaced by the wedge product[11] of one-particle functions $\frac{1}{N!} \bigwedge_{p=1}^N \phi_p$. Let us further introduce a linear operator $\tilde{\mathfrak{A}} : V \to V_\mathfrak{A}$ by

$$\tilde{\mathfrak{A}} f(\vec{\mathbf{x}}) = \frac{1}{\sqrt{N!}} \sum_{P \in \mathcal{S}_N} (-1)^{|P|} f(P\vec{\mathbf{x}}), \tag{2.85}$$

which is equal to the antisymmetrizer (2.79) up to a normalization constant and particularly corresponds to the so-called Slater determinant [170] for a tensor product of one-particle functions $\phi_1, \ldots, \phi_N \in V^{(D)}$, i.e.

$$\tilde{\mathfrak{A}} \bigotimes_{p=1}^N \phi_p = \frac{1}{\sqrt{N!}} \bigwedge_{p=1}^N \phi_p = \frac{1}{\sqrt{N!}} \begin{vmatrix} \phi_1(\mathbf{x}_1) & \cdots & \phi_N(\mathbf{x}_1) \\ \vdots & \ddots & \vdots \\ \phi_1(\mathbf{x}_N) & \cdots & \phi_N(\mathbf{x}_N) \end{vmatrix}. \tag{2.86}$$

Now, if $\mathcal{B}_{V^{(D)}} = \{\mathbf{b}_\mu\}_{\mu \in \Lambda_{(D)}}$ is an orthonormal basis for $V^{(D)}$, then the set

$$\left\{ \tilde{\mathfrak{A}} \vec{\mathbf{b}}_{P\vec{\mu}} : \mu \in \Lambda, \mu_1 < \cdots < \mu_N \right\} \tag{2.87}$$

is an orthonormal basis for the antisymmetric subspace $V_\mathfrak{A}$. Analogously, the tensor product space $V^{(D,N)} = \bigotimes_{p=1}^N V^{(D)}$ gets replaced by the antisymmetric tensor product space $V_\mathfrak{A}^{(D,N)} = \bigwedge_{p=1}^N V^{(D)}$.

---

[11] The wedge product is an antisymmetric variant of the tensor product and is also known as exterior product.

2.5.1 Symmetric and antisymmetric general sparse grid spaces

In the following we consider symmetric and antisymmetric subspaces of the general sparse grid spaces introduced in Section 2.3.2.

Let $f \in \mathcal{S}'((\mathbb{R}^D)^N)$, $h : \mathbb{R} \to \mathbb{C}$ and $\Omega \subset (\mathbb{R}^D)^N$ such that $\int_\Omega h((P\vec{\mathbf{k}})^T \vec{\mathbf{x}}) f(\vec{\mathbf{x}}) \, d\vec{\mathbf{x}} < \infty$ and that Ω is invariant under the permutations of particle coordinates in accordance with (2.77), i.e.

$$\Omega = P\Omega, \quad \forall P \in \mathcal{S}_N.$$

Then, the equation

$$\int_\Omega h((P\vec{\mathbf{k}})^T \vec{\mathbf{x}}) f(\vec{\mathbf{x}}) \, d\vec{\mathbf{x}} = \int_\Omega h(\vec{\mathbf{k}}^T \underbrace{P^{-1}\vec{\mathbf{x}}}_{=\vec{\mathbf{y}}}) f(\vec{\mathbf{x}}) \, d\vec{\mathbf{x}} = \int_\Omega h(\vec{\mathbf{k}}^T \vec{\mathbf{y}}) f(P\vec{\mathbf{y}}) \, d\vec{\mathbf{y}}$$

holds for all permutations $P \in \mathcal{S}_N$. Thus, with $h(t) = e^{-it}$, it easily follows that the Fourier transform $\hat{f}$ exhibits the same symmetry with respect to permutations of particle coordinates as the N-particle function f itself. For example for symmetric $f \in \mathcal{S}'((\mathbb{R}^D)^N)$ the relation $\hat{f}(P\vec{\mathbf{k}}) = \hat{f}(\vec{\mathbf{k}})$ holds for all $P \in \mathcal{S}_N$ and for antisymmetric $f \in \mathcal{S}'((\mathbb{R}^D)^N)$ the relation $\hat{f}(P\vec{\mathbf{k}}) = (-1)^{|P|} \hat{f}(\vec{\mathbf{k}})$ holds for all $P \in \mathcal{S}_N$, again with the usual distributive meaning. Let us remark that for a symmetric or antisymmetric N-particle function f the relation $\|f(P\cdot)\|_{\mathcal{H}^{t,r}_{\text{mix}}} = \|f(\cdot)\|_{\mathcal{H}^{t,r}_{\text{mix}}}$ holds for all $P \in \mathcal{S}_N$; compare Definition 2.2. Note further that a general hyperbolic cross $\mathcal{K}^T_K$, as introduced in Definition 2.3, and also a general hyperbolic cross index set $\mathcal{I}^T_L$, as in Definition 2.4, are invariant under permutations of particles and thus the associated projections $f_{\mathcal{K}^T_K}$ in Lemma 2.1, $f_{\mathcal{I}^T_L}$ in Lemma 2.4 and $f_{\mathcal{I}^T_L, \mathcal{J}^R_J}$ in Lemma 2.5 are symmetric or antisymmetric if f is symmetric or antisymmetric, respectively. Hence, in the framework of Lemma 2.1, Lemma 2.4 and Lemma 2.5 we only have to consider the respective symmetric or antisymmetric subspaces. For example, analogously to relation (2.51) the relation

$$\inf_{\tilde{f}^\mathfrak{A} \in \mathfrak{A}(V^{T;R}_{L;J})} \|f^\mathfrak{A} - \tilde{f}^\mathfrak{A}\|_{\mathcal{H}^{t',r'}_{\text{mix}}} \leq \|f^\mathfrak{A} - f^\mathfrak{A}_{\mathcal{I}^T_L, \mathcal{J}^R_J}\|_{\mathcal{H}^{t',r'}_{\text{mix}}}$$

holds for an antisymmetric N-particle function $f^\mathfrak{A} \in \mathfrak{A}(\mathcal{H}^{t,r,\hat{t},\hat{r}}_{\text{mix}})$ and the relation

$$\inf_{\tilde{f}^\mathfrak{S} \in \mathfrak{S}(V^{T;R}_{L;J})} \|f^\mathfrak{S} - \tilde{f}^\mathfrak{S}\|_{\mathcal{H}^{t',r'}_{\text{mix}}} \leq \|f^\mathfrak{S} - f^\mathfrak{S}_{\mathcal{I}^T_L, \mathcal{J}^R_J}\|_{\mathcal{H}^{t',r'}_{\text{mix}}}$$

holds for a symmetric N-particle function $f^\mathfrak{S} \in \mathfrak{S}(\mathcal{H}^{t,r,\hat{t},\hat{r}}_{\text{mix}})$, where $f^\mathfrak{A}_{\mathcal{I}^T_L, \mathcal{J}^R_J} \in \mathfrak{A}(V^{T;R}_{L;J})$, $f^\mathfrak{S}_{\mathcal{I}^T_L, \mathcal{J}^R_J} \in \mathfrak{S}(V^{T;R}_{L;J})$ and $V^{T;R}_{L;J}$ according to (2.50). In this way, an upper bound estimate for the best approximation error of a symmetric or antisymmetric function in $\mathcal{H}^{t,r,\hat{t},\hat{r}}_{\text{mix}}$ measured in the $\mathcal{H}^{t',r'}_{\text{mix}}$-norm is then again given with relation (2.51) by equation (2.42) together with Lemma 2.4 and Lemma 2.5. Thus, with respect to the achieved accuracies, the order in K and J does not change when we switch to the antisymmetric or symmetric case. Additionally, the involved order constants do not change. In the antisymmetric and symmetric case the same error estimates as in the classical case of Lemma 2.4 and Lemma 2.5 hold.

Note that according to (2.50) the space $V^{T;R}_{L;J}$ is generated by the set $\mathcal{B}_{V^{T;R}_{L;J}}$ of tensor product functions given in (2.46). Thus, a possible generating system for the antisymmetric subspace $V^{T;R}_{L;J;(\mathfrak{A})} := \mathfrak{A}(V^{T;R}_{L;J})$ is given similar to (2.82) with the help of the index

2 Sobolev Spaces for Many-Particle Functions

set
$$\mathcal{I}_{L;J;(\mathfrak{A})}^{T;R} := \left\{ (\vec{l},\vec{\mathbf{j}}) \in \mathcal{I}_{L;J}^{T;R} : (l_1,\mathbf{j}_1) < \cdots < (l_N,\mathbf{j}_N) \right\} \tag{2.88}$$

by
$$\mathcal{B}_{V_{L;J;(\mathfrak{A})}^{T;R}} = \left\{ \mathfrak{A}\phi_{\vec{l}\vec{\mathbf{j}}} \right\}_{(\vec{l},\vec{\mathbf{j}}) \in \mathcal{I}_{L;J;(\mathfrak{A})}^{T;R}} \tag{2.89}$$

and a possible generating system for the symmetric subspace $V_{L;J;(\mathfrak{S})}^{T;R} := \mathfrak{S}(V_{L;J}^{T;R})$ is given analogously to (2.81) with the help of the index set

$$\mathcal{I}_{L;J;(\mathfrak{S})}^{T;R} := \left\{ (\vec{l},\vec{\mathbf{j}}) \in \mathcal{I}_{L;J}^{T;R} : (l_1,\mathbf{j}_1) \leq \cdots \leq (l_N,\mathbf{j}_N) \right\}$$

by
$$\mathcal{B}_{V_{L;J;(\mathfrak{S})}^{T;R}} = \left\{ \mathfrak{S}\phi_{\vec{l}\vec{\mathbf{j}}} \right\}_{(\vec{l},\vec{\mathbf{j}}) \in \mathcal{I}_{L;J;(\mathfrak{S})}^{T;R}} ,$$

where the relations $<$ and $\leq$ are defined for the index pair

$$\mathbf{I}_p := (l_p, \mathbf{j}_p) = (l_p, j_{p,(1)}, \ldots, j_{p,(D)})$$

as

$$\mathbf{I}_p < \mathbf{I}_q :\Leftrightarrow \exists \alpha \in \{1,\ldots,D+1\} : I_{p,(\alpha)} < I_{q,(\alpha)} \land \forall \beta \in \{1,\ldots,\alpha-1\} : I_{p,(\beta)} \leq I_{q,(\beta)}$$

and

$$\mathbf{I}_p \leq \mathbf{I}_q :\Leftrightarrow \exists \alpha \in \{1,\ldots,D+1\} : I_{p,(\alpha)} \leq I_{q,(\alpha)} \land \forall \beta \in \{1,\ldots,\alpha-1\} : I_{p,(\beta)} \leq I_{q,(\beta)},$$

respectively.

For the antisymmetric case we give an upper estimate for $|\mathcal{I}_{L;J;(\mathfrak{A})}^{T;R}|$ according to (2.84) by the following lemma:

Lemma 2.10. *For $T < 1$, $L \in \mathbb{N}$, $R < 1$, $J \in \mathbb{N}$ it holds*

$$|\mathcal{I}_{L;J;(\mathfrak{A})}^{T;R}| \leq \frac{1}{N!} |\mathcal{I}_{L;J}^{T;R}|.$$

We see that the order for the dimension of the different spaces with respect to L and J stays the same for all different cases of T and R as in the classical case of Lemma 2.7 without antisymmetry. However, the constant is now reduced by the factor $1/N!$. This is a substantial improvement which allows the treatment larger particle numbers in the antisymmetric case.

2.5.2 Particle-wise splitting of symmetric and antisymmetric spaces

In the following we discuss the application to symmetric and antisymmetric many-particle spaces of a particle-wise subspace splitting in the framework of many-particle spaces of finite-order weights as considered in Section 2.4.

2.5 Symmetric and antisymmetric many-particle spaces

Particle-wise subspace splitting of antisymmetric many-particle spaces

Let $f \in V_{\mathfrak{A}}^{(D,N)} \subset V^{(D,N)}$. As in Section 2.4.1, $V^{(D)}$ denotes a separable Hilbert space of D-dimensional functions, $V^{(D,N)}$ the corresponding N-particle space $V^{(D,N)} = \bigotimes_{p=1}^{N} V^{(D)}$ and $\mathcal{N}$ denotes the set $\mathcal{N} = \{1, \ldots, N\}$.

First, let us shortly discuss the case of a first order particle-wise decomposition of the antisymmetric N-particle function f, where the decomposition is given with respect to N normalized one-particle functions $\{g\}_{p \in \mathcal{N}}$ according to (2.68). In particular, the functions F_u are generally not in antisymmetric subspaces. To obtain a subspace splitting of the antisymmetric tensor product space $V_{\mathfrak{A}}^{(D,N)} = \bigwedge_{p=1}^{N} V^{(D)}$ analogous to (2.58), (2.66) and (2.63), we may employ the antisymmetrizer, i.e.

$$V_{\mathfrak{A}}^{(D,N)} = \mathfrak{A}\left(\bigoplus_{\mu=0}^{N} \bigoplus_{u \subset \mathcal{N}, |u|=\mu} \bigotimes_{p=1}^{N} \mathcal{W}_{u,(p)} \right)$$

$$= \sum_{\mu=0}^{N} \mathfrak{A}\left(\bigoplus_{u \subset \mathcal{N}, |u|=\mu} \bigotimes_{p=1}^{N} \mathcal{W}_{u,(p)} \right)$$

with $\mathcal{W}_{u,(p)}$ as in (2.59). Accordingly, a function $f \in V_{\mathfrak{A}}^{(D,N)}$ is decomposed as

$$f(\vec{\mathbf{x}}) = \mathfrak{A}\left(\sum_{u \subset \mathcal{N}} F_u \right)(\vec{\mathbf{x}})$$

$$= \sum_{u \subset \mathcal{N}} \mathfrak{A}(F_u)(\vec{\mathbf{x}})$$

$$= \sum_{u \subset \mathcal{N}} \mathfrak{A}\left(\vec{\mathbf{y}} \mapsto f_u(\vec{\mathbf{y}}_u) \prod_{p \in \mathcal{N} \setminus u} g_p(\mathbf{y}_p) \right)(\vec{\mathbf{x}}),$$

where $F_u \in \mathcal{W}_u$ and $f_u \in \bigotimes_{p \in u} W_p$. Note that particularly in the case of $g_1 = \cdots = g_N$ most low-particle functions f_u vanish, since then $\mathfrak{A}(F_u) = 0$ for $N - |u| > 1$.

To obtain a direct sum decomposition of the antisymmetric space $V_{\mathfrak{A}}^{(D,N)}$ we consider a higher order splitting as in Section 2.4.1. Let the subspace U be spanned by linear independent normalized one-particle functions $g_1, \ldots, g_N \in V^{(D)}$, i.e. $N_U = N$ in Section 2.4.1, and let $W \subset V^{(D)}$ such that $V^{(D)} = U \oplus W$. Then, corresponding to (2.72) with $\mathcal{U}_{u_1, \ldots, u_N, (p)}$ from (2.71) we obtain a splitting in the form

$$V_{\mathfrak{A}}^{(D,N)} = \mathfrak{A}\left(\bigotimes_{p=1}^{N} (U \oplus W) \right)$$

$$= \mathfrak{A}\left(\bigoplus_{\mu=0}^{N} \bigoplus_{u_1, \ldots, u_N \subset \mathcal{N}, \forall \nu \neq \nu': u_\nu \cap u_{\nu'} = \emptyset, |\mathcal{N} \setminus \bigcup_{\nu=1}^{N} u_\nu| = \mu} \mathcal{U}_{u_1, \ldots, u_N, (p)} \right)$$

$$= \bigoplus_{\mu=0}^{N} \mathfrak{A}\left(\bigoplus_{u_1, \ldots, u_N \subset \mathcal{N}, \forall \nu \neq \nu': u_\nu \cap u_{\nu'} = \emptyset, |\mathcal{N} \setminus \bigcup_{\nu=1}^{N} u_\nu| = \mu} \mathcal{U}_{u_1, \ldots, u_N, (p)} \right)$$

$$= \bigoplus_{\mu=0}^{N} \underbrace{\bigoplus_{u \subset \mathcal{N}, |u|=\mu} \mathcal{W}_u^{\mathfrak{A}}}_{=: \mathcal{V}_\mu^{\mathfrak{A}}}, \tag{2.90}$$

where

$$\mathcal{W}_u^{\mathfrak{A}} := \mathfrak{A}\left(\bigotimes_{p=1}^{N} \mathcal{W}_{u,(p)}^{\mathfrak{A}}\right), \quad \mathcal{W}_{u,(p)}^{\mathfrak{A}} := \begin{cases} \mathrm{span}\{g_p\} & \text{for } p \in \mathcal{N} \setminus u, \\ W & \text{for } p \in u. \end{cases} \qquad (2.91)$$

Correspondingly, with the help of appropriate linear projections a function $f \in V_{\mathfrak{A}}^{(D,N)}$ can be decomposed as

$$f = \mathfrak{A}\left(\sum_{u \subset \mathcal{N}} F_u\right) = \sum_{u \subset \mathcal{N}} \mathfrak{A}(F_u)(\vec{x}) = \sum_{u \subset \mathcal{N}} \mathfrak{A}\left(\vec{y} \mapsto f_u(\vec{y}_u) \prod_{p \in \mathcal{N} \setminus u} g_p(y_p)\right), \qquad (2.92)$$

such that $\mathfrak{A}F_u \in \mathcal{W}_u^{\mathfrak{A}}$ and $f_u \in \bigotimes_{p \in u} W$. Moreover, in the case of an orthogonal direct sum $U \oplus W$ and orthonormal $\{g_p\}_{p \in \mathcal{N}}$, the orthogonality relation $\langle F_u, F_v \rangle_{V^{(D,N)}} = 0$ holds for all $u \neq v$. Let us remark that such a type of a particle-wise decomposition of an antisymmetric N-particle function is also known in quantum chemistry as *successive partial orthogonalizations* method [164], where particles are associated with electrons. Regarding the choice of the one-particle functions $\{g_p\}_{p \in \mathcal{N}}$ to gain a fast decay in the size of the terms $\|f_u\|_{\mathcal{L}^2}$ with $\mu = |u|$ one might minimize the approximation error of $\|f - \mathfrak{A}F_\emptyset\|_{\mathcal{L}^2}$, i.e. one might employ one-particle functions which correspond to a best (antisymmetric) rank-1 approximation with respect to the $\mathcal{L}^2$-norm

$$\|f - \mathfrak{A}F_\emptyset\|_{\mathcal{L}^2} = \|f - f_\emptyset \mathfrak{A}\bigotimes_{p=1}^{N} g_p\|_{\mathcal{L}^2} = \min_{\tilde{g}_1,\ldots,\tilde{g}_N} \|f - \bigwedge_{p=1}^{N} \tilde{g}_p\|_{\mathcal{L}^2}. \qquad (2.93)$$

Furthermore, let us consider the case that the one-particle space $V^{(D)}$ is spanned by a finite basis set $\mathcal{B}_{V^{(D)}} = \{\mathbf{b}_\nu\}_{\nu \in \Lambda_{(D)}}$, with $\Lambda_{(D)} = \{1, \ldots, N+m\}$ and that we choose $g_p = \mathbf{b}_p$, for $p = 1, \ldots, N$. Then the spaces $\mathcal{V}_\mu^{\mathfrak{A}}$ in (2.90) are spanned by

$$\mathcal{V}_\mu^{\mathfrak{A}} = \mathrm{span}\left\{\mathfrak{A}\bigotimes_{p \in \mathcal{N} \setminus u} g_p \otimes \bigotimes_{q=1}^{\mu} \mathbf{b}_{\nu_q} : u \subset \mathcal{N}, |u| = \mu, N < \nu_1 < \cdots < \nu_\mu \leq N+m\right\}$$

for $\mu = 0, \ldots, N$, where for its numbers of degrees of freedom follows that

$$|\mathcal{V}_\mu^{\mathfrak{A}}| = \binom{N}{N-\mu}\binom{m}{\mu} = \binom{N}{\mu}\binom{m}{\mu} \leq \frac{N^\mu m^\mu}{\mu!\mu!}. \qquad (2.94)$$

Let us remark that the use of finite-order weights of order q in the framework of the subspace splitting (2.90) leads to the restriction to a subspace

$$\bigoplus_{\mu=0}^{q} \bigoplus_{u \subset \mathcal{N}, |u|=\mu} \mathcal{W}_u^{\mathfrak{A}} = \bigoplus_{\mu=0}^{q} \mathcal{V}_\mu^{\mathfrak{A}} \subset V_{\mathfrak{A}}^{(D,N)} \qquad (2.95)$$

Note finally that the subspace splitting (2.90) and its restriction (2.95) are related to the so-called configuration interaction (CI) approaches in quantum chemistry [170]. For more details see also Section 4.3.2.

2.5 Symmetric and antisymmetric many-particle spaces

Particle-wise decomposition of symmetric functions

For $f \in V_\mathfrak{S}^{(D,N)} \subset V^{(D,N)}$, let us shortly discuss the case of a first order particle-wise decomposition, which is given with respect to the N normalized one-particle functions $\{g\}_{p \in \mathcal{N}}$ from (2.68). Analogous to the antisymmetric case, the functions F_u are in general not symmetric. To obtain a subspace splitting of $V_\mathfrak{S}^{(D,N)}$, similar to (2.58), (2.66) and (2.63), we may employ the symmetrizer, i.e.

$$V_\mathfrak{S}^{(D,N)} = \mathfrak{S}\left(\bigoplus_{\mu=0}^{N} \bigoplus_{u \subset \mathcal{N}, |u|=\mu} \bigotimes_{p=1}^{N} \mathcal{W}_{u,(p)}\right) = \sum_{\mu=0}^{N} \mathfrak{S}\left(\bigoplus_{u \subset \mathcal{N}, |u|=\mu} \bigotimes_{p=1}^{N} \mathcal{W}_{u,(p)}\right).$$

Accordingly, a function $f \in V_\mathfrak{S}^{(D,N)}$ is decomposed as

$$f = \mathfrak{S}\left(\sum_{u \subset \mathcal{N}} F_u\right) = \sum_{\mu=0}^{N} \sum_{u \subset \mathcal{N}, |u|=\mu} \mathfrak{S}\left(\vec{y} \mapsto f_u(\vec{y}_u) \prod_{p \in \mathcal{N} \setminus u} g_p(\mathbf{y}_p)\right),$$

where $F_u \in \mathcal{W}_u$ and $f_u \in \bigotimes_{p \in u} W_p$.

Let us now consider the case that the normalized one-particle functions are all equal, i.e. $g := g_1 = \cdots = g_N$, $U := \text{span}\{g\}$ and $V^{(D)} = U \oplus W$. Then the relations

$$V_\mathfrak{S}^{(D,N)} = \mathfrak{S}\left(\bigoplus_{\mu=0}^{N} \bigoplus_{u \subset \mathcal{N}, |u|=\mu} \mathcal{W}_u\right) = \bigoplus_{\mu=0}^{N} \underbrace{\mathfrak{S}\left(\bigotimes_{p=1}^{N-\mu} U \otimes \bigotimes_{p=1}^{\mu} W\right)}_{=: \mathcal{V}_\mu^\mathfrak{S}} \tag{2.96}$$

and

$$f(\vec{x}) = \sum_{\mu=0}^{N} \binom{N}{\mu} \mathfrak{S}(F_{\{1,\ldots,\mu\}})(\vec{x}) = \sum_{\mu=0}^{N} \binom{N}{\mu} \mathfrak{S}\left(f_{\{1,\ldots,\mu\}} \otimes \bigotimes_{p=1}^{N-\mu} g\right)(\vec{x})$$

hold. Note particularly that there are only $N+1$ different functions $f_\emptyset, f_{\{1\}}, \ldots, f_{\{1,\ldots,N\}}$. Moreover, if the one-particle space $V^{(D)}$ is spanned by a finite basis set

$$\mathcal{B}_{V^{(D)}} = \{\mathbf{b}_\nu\}_{\nu \in \Lambda_{(D)}} \quad \text{with } \Lambda_{(D)} = \{1, \ldots, 1+m\}$$

and if we choose $g = \mathbf{b}_1$, then the spaces $\mathcal{V}_\mu^\mathfrak{S}$ are spanned by

$$\mathcal{V}_\mu^\mathfrak{S} = \text{span}\left\{\mathfrak{S} \bigotimes_{p=1}^{N-\mu} g \otimes \bigotimes_{q=1}^{\mu} \mathbf{b}_{\nu_q} : 1 < \nu_1 \leq \cdots \leq \nu_\mu \leq m+1\right\}$$

for $\mu = 0, \ldots, N$, where for its cardinality numbers follows that

$$|\mathcal{V}_\mu^\mathfrak{S}| = \binom{m+N-1}{\mu} \leq \frac{(m+N-1)^\mu}{\mu!}.$$

Similar to the antisymmetric case, the use of finite-order weights of order q in the framework of the subspace splitting (2.96) leads to the restriction to a subspace

$$\mathfrak{S}\left(\bigoplus_{\mu=0}^{q} \bigoplus_{u \subset \mathcal{N}, |u|=\mu} \mathcal{W}_u\right) = \bigoplus_{\mu=0}^{q} \mathcal{V}_\mu^\mathfrak{S} \subset V_\mathfrak{S}^{(D,N)}.$$

2.6 Summary

We generalized the optimized general tensor product approximation spaces introduced by Griebel and Knapek [72] for the case of one-dimensional particles with periodic boundary conditions, i.e. $\mathcal{H}_{\text{mix}}^{t,r}((\mathbb{T}^1)^N)$, to the case of N D-dimensional particles with decay conditions, i.e. $\mathcal{H}_{\text{mix}}^{t,r}((\mathbb{R}^D)^N)$ and $\mathcal{H}_{\text{mix}}^{t,r;\hat{t},\hat{r}}((\mathbb{R}^D)^N)$. We further gave estimates for the approximation and complexity orders. Here, we particularly balanced the error related to the truncation in Fourier space and the error related to the truncation in real space. Due to possibly large constants involved in the approximation and complexity order estimates the generalized sparse grid discretization scheme is only practical for a moderate number of low-dimensional particles even if no exponentially or logarithmically dependent terms with respect to the discretization parameter are present. Nevertheless, the discretization scheme can be applied in the framework of many-particle spaces with finite-order weights of sufficiently low order. To this end, we introduced and considered appropriately weighted spaces which are related to certain first or higher order particle-wise subspace splittings. Moreover, we discussed the application of symmetry conditions. Here, it particularly turned out that the successive partial orthogonalization method for analysing wave functions in quantum chemistry can be identified with a certain higher-order particle-wise subspace splitting approach in the framework of finite-order weighted spaces.

3 Electronic Schrödinger Equation

In quantum chemistry the main interest is to find an approximation of the so-called ground-state of an atom, molecule or ion. All statements that are made about a quantum mechanical system can be derived from the so-called state function (or wave function) which is given by the solution of the Schrödinger equation. The Schrödinger equation is a high-dimensional eigenvalue problem and the ground-state is the eigenstate, which is an eigenfunction of the Hamilton operator which corresponds to the lowest eigenvalue. In this chapter we introduce the Born-Oppenheimer approximation of the Schrödinger equation which is central to quantum chemistry and discuss some important properties of its solution wave functions which are of special interest for this thesis.

3.1 Born-Oppenheimer approximation

Let us consider the time-independent (or stationary) Schrödinger equation

$$H\Psi = E\Psi,$$

where in atomic units the non-relativistic Hamilton operator H associated with a system of N electrons and N_{nuc} nuclei is given by

$$H := -\frac{1}{2}\sum_{p=1}^{N}\Delta_{\mathbf{x}_p} - \sum_{p=1}^{N}\sum_{q=1}^{N_{\text{nuc}}}Z_q v(\mathbf{x}_p - \mathbf{R}_q) + \sum_{p=1}^{N}\sum_{p'>p}^{N} v(\mathbf{x}_p - \mathbf{x}_{p'})$$
$$+ \sum_{q=1}^{N_{\text{nuc}}}\sum_{q'>q}^{N_{\text{nuc}}} Z_q Z_{q'} v(\mathbf{R}_q - \mathbf{R}_{q'}) - \frac{1}{2}\sum_{q=1}^{N_{\text{nuc}}}\frac{1}{2M_q}\Delta_{\mathbf{R}_q}, \quad (3.1)$$

with the Coulomb interaction potential $v(\mathbf{r}) = \frac{1}{|\mathbf{r}|_2}$. Note that we use throughout this thesis atomic units for all given quantities. Here, M_q and Z_q denote the mass and the atomic number of the q-th nucleus. Furthermore, $\mathbf{x}_p \in \mathbb{R}^3$ denotes the position of the p-th electron and $\mathbf{R}_q \in \mathbb{R}^3$ denotes the position of the q-th nucleus. The operators $\Delta_{\mathbf{x}_j}$ and $\Delta_{\mathbf{R}_q}$ stand for the Laplace operator with respect to the coordinates of the p-th electron and the q-th nucleus, respectively. The first term of (3.1) corresponds to the kinetic energy of the electrons, the second term corresponds to the potential energy of the interactions between electrons and nuclei, the third term corresponds to the potential energy of the repulsion between the electrons, the fourth term corresponds to the potential energy of the repulsion between the nuclei and the last term corresponds to the kinetic energy of the nuclei. Accordingly, the Hamilton operator (3.1) in abbreviated form reads

$$H = T_e + V_{en} + V_{ee} + V_{nn} + T_n. \quad (3.2)$$

In this thesis we mainly consider the so-called Born-Oppenheimer approximation. In order to reduce the number of variables we assume that the electrons move in the field

3 Electronic Schrödinger Equation

given by the fixed nuclei. This assumption is based on the large difference in masses of the electrons and the nuclei. Thus, in (3.2) the kinetic energy operator T_n of the nuclei is neglected and the operator V_{nn}, which corresponds to the potential energy of the interaction of the nuclei, is considered to be constant. Note that in this way the constant term V_{nn} has no effect on the eigenfunctions of the operator and just adds on the eigenvalues. The resulting so-called electronic Hamilton operator for N_{nuc} given fixed point charges reads as

$$H_e^{(\vec{R},\vec{Z})} := T_e + V_{en}^{(\vec{R},\vec{Z})} + V_{ee} \qquad (3.3)$$

and the corresponding Schrödinger equation

$$H_e^{(\vec{R},\vec{Z})} \Psi_e^{(\vec{R},\vec{Z})} = E_e^{(\vec{R},\vec{Z})} \Psi_e^{(\vec{R},\vec{Z})}, \qquad (3.4)$$

which is also called the electronic Schrödinger equation. Accordingly, the eigenvalues $E_e^{(\vec{R},\vec{Z})}$ and the corresponding eigenfunctions $\Psi_e^{(\vec{R},\vec{Z})}$ are called electronic energies and electronic wave functions, respectively. The so-called total energies within the Born-Oppenheimer approximation are given by

$$E_{tot}^{(\vec{R},\vec{Z})} := E_e^{(\vec{R},\vec{Z})} + V_{nn}^{(\vec{R},\vec{Z})}. \qquad (3.5)$$

Note that a superscript $(\vec{R}, \vec{Z})$ on a term denotes parametric dependence on the coordinates $\vec{R} := (\mathbf{R}_1, \ldots, \mathbf{R}_{N_{\text{nuc}}})$ and the charges $\vec{Z} := (Z_1, \ldots, Z_{N_{\text{nuc}}})$. Note further that the Born-Oppenheimer approximation can be used to compute the total energy (3.5) given to a fixed arrangement of nuclei $(\vec{R}, \vec{Z})$. Thus an approximation of the ground-state can be found by varying the total energy (3.5) with respect to the coordinates of the nuclei $\vec{R}$. Therefore, we consider from now on only the electronic Hamilton operator (3.3), the electronic Schrödinger equation (3.4) and the total energy (3.5). In particular we drop the superscript $(\vec{R}, \vec{Z})$ and the subscript e in the following, i.e. we denote the electronic Hamilton operator, the electronic wave function and the electronic energy by H, Ψ and E, respectively.

3.2 Antisymmetry principle

So far we did not include the spin of each electron, since the non-relativistic electronic Hamiltonian (3.3) does not depend on it. However, in general, an electronic wave function depends not only on the positions $\mathbf{x}_i \in \mathbb{R}^3$ of the electrons but also on their associated spin coordinates $s_p \in \{+\frac{1}{2}, -\frac{1}{2}\}$. Thus, electronic wave functions are defined by[1]

$$\Psi : (\mathbb{R}^3)^N \times \{+\tfrac{1}{2}, -\tfrac{1}{2}\}^N \to \mathbb{R} : (\vec{\mathbf{x}}, \vec{s}) \mapsto \Psi(\vec{\mathbf{x}}, \vec{s}) \qquad (3.6)$$

with the $3N$-dimensional spatial coordinate $\vec{\mathbf{x}} := (\mathbf{x}_1, \ldots, \mathbf{x}_N)$ with $\mathbf{x}_p \in \mathbb{R}^3$ and with the N-dimensional spin coordinate $\vec{s} := (s_1, \ldots, s_N)$ with $s_p \in \{+\frac{1}{2}, -\frac{1}{2}\}$. Furthermore, the following two assumptions from quantum mechanics have to be taken into account: First, elementary particles are indistinguishable from each other (fundamental principle

---
[1] Note that in general, an electronic wave function is defined as $\Psi : (\mathbb{R}^3)^N \times \{+\frac{1}{2}, -\frac{1}{2}\}^N \to \mathbb{C}$. However, due to the definition of the electronic Hamilton operator (3.3), it is enough to consider wave functions as in (3.6) only.

3.2 Antisymmetry principle

of quantum mechanics). Second, no two electrons may occupy the same quantum state simultaneously (Pauli exclusion principle). Therefore, an electron wave function has to obey the antisymmetry principle which states that an N-electron wave function (3.6) is antisymmetric with respect to an arbitrary simultaneous permutation $P \in \mathcal{S}_N$ of the electron positions and spin variables, i.e. it fulfills

$$\Psi(P\vec{\mathbf{x}}, P\vec{s}) = (-1)^{|P|}\Psi(\vec{\mathbf{x}}, \vec{s}), \quad \forall P \in \mathcal{S}_N. \tag{3.7}$$

Here, in accordance with Section 2.5, $\mathcal{S}_N$ denotes the symmetric group of degree N. Furthermore, the permutation P is a mapping $P : \{1, \ldots, N\} \to \{1, \ldots, N\}$ which translates to a permutation of the corresponding numbering of electrons, i.e. $P(\mathbf{x}_1, \ldots, \mathbf{x}_N) := (\mathbf{x}_{P(1)}, \ldots, \mathbf{x}_{P(N)})$ and $P(s_1, \ldots, s_N) := (s_{P(1)}, \ldots, s_{P(N)})$, and the expression $(-1)^{|P|}$ is equal to the sign, or parity, of the permutation P.

Let us consider an electronic wave function Ψ, i.e. an eigenfunction Ψ of the electronic Hamiltonian (3.3) which obeys the antisymmetry principle (3.7). For a given spin vector $\vec{s} \in \{+\frac{1}{2}, -\frac{1}{2}\}^N$ we define the associated spatial component of the electronic wave function Ψ by

$$\Psi_{\vec{s}} : (\mathbb{R}^3)^N \to \mathbb{R} : \vec{\mathbf{x}} \to \Psi(\vec{\mathbf{x}}, \vec{s})$$

and define its associated total spin projection[2] by

$$M_S^{\vec{s}} := \sum_{p=1}^{N} s_p. \tag{3.8}$$

Then, since there are 2^N possible different spin distributions $\vec{s}$, the electronic Schrödinger equation, i.e. the eigenvalue problem

$$H\Psi = E\Psi,$$
$$\Psi(P\vec{\mathbf{x}}, P\vec{s}) = (-1)^{|P|}\Psi(\vec{\mathbf{x}}, \vec{s}), \quad \forall P \in \mathcal{S}_N \tag{3.9}$$

decouples into 2^N eigenvalue problems for the 2^N associated spatial components $\Psi_{\vec{s}}$. Here, for given $\vec{s}$ the spatial part $\Psi_{\vec{s}}$ obeys the partially antisymmetry condition

$$\Psi_{\vec{s}}(P\vec{\mathbf{x}}) = (-1)^{|P|}\Psi_{\vec{s}}(P\vec{\mathbf{x}}), \quad \forall P \in \mathcal{S}_{\vec{s}} := \{P \in \mathcal{S}_N : P\vec{s} = \vec{s}\}, \tag{3.10}$$

where $\mathcal{S}_{\vec{s}} \subset \mathcal{S}_N$ is a permutation group. In particular, for the spatial components the minimal eigenvalue of all eigenvalue problems is equal to the minimal eigenvalue of the full eigenvalue problem (3.9). Moreover, the eigenfunctions of the full system (3.9) can be composed of the eigenfunctions of the eigenvalue problems for the spatial parts. For further details see also [192, 196]. Note that although there are 2^N possible different spin distributions $\vec{s}$ for an N-electron wave function Ψ, there are only $N+1$ different possible values for the total spin projection (3.8). Furthermore, under the equivalence relation

$$\vec{s} \sim_S \vec{s}' :\Leftrightarrow M_S^{\vec{s}} = M_S^{\vec{s}'}$$

the set of all possible equivalence classes of $\{+\frac{1}{2}, -\frac{1}{2}\}^N$ is equal to the set of all possible equivalence classes of $\{+\frac{1}{2}, -\frac{1}{2}\}^N$ under the equivalence relation

$$\vec{s} \sim_{\mathcal{S}_N} \vec{s}' :\Leftrightarrow \exists P \in \mathcal{S}_N : P\vec{s} = \vec{s}',$$

---

[2] In quantum chemistry M_S usually denotes the z component of the total spin of an N-electron state. For a further reading see for example the textbooks [85, 170].

3 Electronic Schrödinger Equation

which is related to the partially antisymmetry conditions. Thus, it is sufficient to consider $N+1$ eigenvalue problems which are associated with $N+1$ different class representative spin vectors with a total spin projection of values $-\frac{N}{2},\ldots,-1,0,1,\ldots,\frac{N}{2}$ for even N and a total spin projection of values $-\frac{N}{2},\ldots,-\frac{1}{2},\frac{1}{2},\ldots,\frac{N}{2}$ for odd N. To this end, we choose the $N+1$ different class representative spin vectors

$$\vec{s}^{(N,M_S)} = (s_1^{(N,M_S)},\ldots,s_N^{(N,M_S)}) \in \{+\tfrac{1}{2},-\tfrac{1}{2}\}^N,$$

where the first $\frac{N}{2}+M_S$ electrons possess spin $+\frac{1}{2}$ and the remaining $\frac{N}{2}-M_S$ electrons possess spin $-\frac{1}{2}$, i.e.

$$s_j^{(N,M_S)} := \begin{cases} +\frac{1}{2} & \text{for } j \leq \frac{N}{2}+M_S, \\ -\frac{1}{2} & \text{for } j > \frac{N}{2}+M_S, \end{cases} \quad (3.11)$$

where $M_S \in \{-\frac{N}{2},\ldots,-1,0,1,\ldots,\frac{N}{2}\}$ for even N and $M_S \in \{-\frac{N}{2},\ldots,-\frac{1}{2},\frac{1}{2},\ldots,\frac{N}{2}\}$ for odd N. In this way, the total spin projection $M_{\vec{s}^{(N,M_S)}}$ is equal to M_S. Therefore, in the following without loss of generality we only consider eigenvalue problems

$$H\Psi^{(N,M_S)} = E^{(N,M_S)}\Psi^{(N,M_S)},$$
$$\Psi^{(N,M_S)}(P\vec{x}) = (-1)^{|P|}\Psi^{(N,M_S)}(P\vec{x}), \quad \forall P \in \mathcal{S}^{(N,M_S)}, \quad (3.12)$$

which correspond to the $N+1$ different class representative spin distributions $\vec{s}^{(N,M_S)}$ according to (3.11). Here, we set $\mathcal{S}^{(N,M_S)} := \mathcal{S}_{\vec{s}^{(N,M_S)}}$ with $|\mathcal{S}^{(N,M_S)}| = (\frac{N}{2}+M_S)!(\frac{N}{2}-M_S)!$. Note that the eigenfunctions $\Psi^{(N,M_S)}$ are spatial functions

$$\Psi^{(N,M_S)} : (\mathbb{R}^3)^N \to \mathbb{R},$$

where the spin coordinates $\vec{s}^{(N,M_S)}$ impose the partial antisymmetry conditions. Hence, the spatial eigenfunctions $\Psi^{(N,M_S)}$ are in

$$\mathcal{L}^2_{(N,M_S)} := \left(\bigwedge_{p=1}^{N/2+M_S} \mathcal{L}^2(\mathbb{R}^3)\right) \otimes \left(\bigwedge_{p=N/2+M_S+1}^{N} \mathcal{L}^2(\mathbb{R}^3)\right), \quad (3.13)$$

compare Section 2.5. In particular, the full wave function which is given by

$$\Psi : (\mathbb{R}^3)^N \times \{-\tfrac{1}{2},+\tfrac{1}{2}\}^N \to \mathbb{R} : (\vec{x},\vec{s}) \mapsto \frac{1}{N!} \sum_{P \in \mathcal{S}_N} (-1)^P \Psi^{(N,M_S)}(P\vec{x}) \delta_{P\vec{s},\vec{s}^{(N,M_S)}},$$

solves the full eigenvalue problem (3.9) [196].

Furthermore, we label the electrons which possess spin $+\frac{1}{2}$ also by spin-up $\uparrow$ and label the electrons which possess spin $-\frac{1}{2}$ by spin-down $\downarrow$. In this way, for an N-electron state function, we denote the number of spin-up particles by $N_\uparrow$ and the number of spin-down particles by $N_\downarrow$. In particular it holds $N = N_\uparrow + N_\downarrow$ with

$$N_\uparrow = M_S + \frac{N}{2}, \quad N_\downarrow = M_S - \frac{N}{2} \quad (3.14)$$

for the total spin projection $M_S = \frac{1}{2}(N_\uparrow - N_\downarrow)$.

In the case of a spin-independent electronic Hamiltonian operator, it is sufficient to solve the $\lfloor N/2 \rfloor + 1$ eigenvalue problems only, which correspond to spin vectors $\vec{s}^{(N,M_S)}$

with a total spin projection of $0 \leq M_S \leq N/2$, i.e. $0 \leq N_\downarrow \leq N_\uparrow \leq N$. Since in this thesis we do not consider spin operators, we refer to textbooks on quantum chemistry like [85, 170]. However, let us note that besides the spin quantum number M_S, there is another spin quantum number usually denoted by S. The exact eigenfunctions Ψ of a spin-independent electronic Hamilton operator H are also eigenfunctions of the total spin angular momentum operator $\hat{S}\Psi = M_S\Psi$ and its squared-magnitude operator $\hat{S}^2\Psi = S(S+1)\Psi$ since $\hat{S}$ and $\hat{S}^2$ commute with H. For an N-electron state S and M_S describe the total spin and its z component. In this framework states with $S = 0, \frac{1}{2}, 1, \frac{3}{2}, \ldots$ have multiplicity $(2S+1) = 1, 2, 3, 4, \ldots$ and are denoted as singlets ($^1\Psi$), doublets ($^2\Psi$), triplets ($^3\Psi$), quartets ($^4\Psi$), ...; see [170]. Moreover, for the exact N-electron state the spin quantum number S equals $\frac{1}{2}$ times the number of unpaired electrons, i.e. $S = \frac{1}{2}|N_\uparrow - N_\downarrow|$.

3.3 Variational formulation

Note that the methods in this thesis used to compute an approximation of the solution of the electronic Schrödinger equation are based on the variational principle. Thus, we briefly resume the variational formulation of the eigenvalue problem (3.12). Here, due to the kinetic energy operator of the electrons we only consider wave functions in the Sobolev space $\mathcal{H}^1((\mathbb{R}^3)^N) \subset \mathcal{L}^2((\mathbb{R}^3)^N)$. In the following for a shorter notation we write $\mathcal{L}^2$ and $\mathcal{H}^1$ instead of $\mathcal{L}^2((\mathbb{R}^3)^N)$ and $\mathcal{H}^1((\mathbb{R}^3)^N)$. Furthermore, we consider $\mathcal{L}^2$-normed wave functions only, i.e. $\|\Psi\|_{\mathcal{L}^2} = 1$, which obey the partially antisymmetry conditions given in (3.12). Now, let $\mathcal{H}^1_{(N,M_S)} \subset \mathcal{H}^1$ denote the subspace of first-order weakly differentiable partially antisymmetric functions given by[3]

$$\mathcal{H}^1_{(N,M_S)} := \left\{ f \in \mathcal{H}^1 : (-1)^{|P|}f(P\vec{x}) - f(\vec{x}) = 0, \forall P \in \mathcal{S}^{(N,M_S)} \right\} \subset \mathcal{L}^2_{(N,M_S)}.$$

See also Definition (3.13) and Section 2.5. Let us further introduce the linear partial antisymmetrization projection operator $\mathfrak{A}^{(N,M_S)} : \mathcal{L}^2((\mathbb{R}^3)^N) \to \mathcal{L}^2_{(N,M_S)}$ by

$$\begin{aligned}\mathfrak{A}^{(N,M_S)}f(\vec{x}) &:= \left(\mathfrak{A}_{N_\uparrow} \otimes \mathfrak{A}_{N_\downarrow} \right) f(\vec{x}) \\ &= \frac{1}{N_\uparrow! N_\downarrow!} \sum_{P \in \mathcal{S}^{(N,M_S)}} (-1)^{|P|} f(P\vec{x}),\end{aligned} \quad (3.15)$$

where $\mathfrak{A}_{N_\uparrow}$ and $\mathfrak{A}_{N_\downarrow}$ denote the antisymmetric projections according to (2.79) for the subspaces $\mathcal{L}^2((\mathbb{R}^3)^{N_\uparrow})$ and $\mathcal{L}^2((\mathbb{R}^3)^{N_\downarrow})$, respectively. Here, $\mathcal{L}^2_{(N,M_S)} = \mathfrak{A}^{(N,M_S)}(\mathcal{L}^2((\mathbb{R}^3)^N))$ and the numbers $N_\uparrow$ and $N_\downarrow$ are determined by N and M_S given in (3.14).

Now, a function $\Psi^{(N,M_S)} \in \mathcal{H}^1_{(N,M_S)}$ with $\|\Psi^{(N,M_S)}\|_{\mathcal{L}^2} = 1$ is a weak solution of the eigenvalue equation (3.12) with the associated eigenvalue $E^{(N,M_S)}$ if

$$\langle \phi, H\Psi^{(N,M_S)} \rangle_{\mathcal{L}^2} = E^{(N,M_S)} \langle \phi, \Psi^{(N,M_S)} \rangle_{\mathcal{L}^2} \quad (3.16)$$

---

[3] Note that $\mathcal{H}^1_{(N,M_S)}$ may also be introduced as the intersection space $\mathcal{H}^1 \cap \mathcal{L}^2_{(N,M_S)}$ or may also be defined as the closure of

$$\mathcal{S}_{(N,M_S)} := \left\{ f \in \mathcal{S} : (-1)^{|P|} f(\vec{x}) - f(\vec{x}) = 0, \forall P \in \mathcal{S}^{(N,M_S)} \right\}$$

in the space $\mathcal{H}^1$.

for all test functions ϕ in the Sobolev space $\mathcal{H}^1$.[4] It is sufficient to consider only test functions $\phi^{(N,M_S)} \in \mathcal{H}^1_{(N,M_S)}$, since the linear partial antisymmetrization projection operator $\mathfrak{A}^{(N,M_S)}$ and the purely symmetric electronic Hamilton operator H commute [14]. Thus, besides the identity $\mathfrak{A}^{(N,M_S)}\phi^{(N,M_S)} = \phi^{(N,M_S)}$ for $\phi^{(N,M_S)} \in \mathcal{H}^1_{(N,M_S)}$, the identities

$$\langle \phi, H\Psi^{(N,M_S)}\rangle_{\mathcal{L}^2} = \langle \phi, H\mathfrak{A}^{(N,M_S)}\Psi^{(N,M_S)}\rangle_{\mathcal{L}^2} = \langle \mathfrak{A}^{(N,M_S)}\phi, H\Psi^{(N,M_S)}\rangle_{\mathcal{L}^2},$$
$$\langle \phi, \Psi^{(N,M_S)}\rangle_{\mathcal{L}^2} = \langle \phi, \mathfrak{A}^{(N,M_S)}\Psi^{(N,M_S)}\rangle_{\mathcal{L}^2} = \langle \mathfrak{A}^{(N,M_S)}\phi, \Psi^{(N,M_S)}\rangle_{\mathcal{L}^2} \quad (3.17)$$

hold.

The smallest energy with respect to (N, M_S) is given by

$$E_{\min}^{(N,M_S)} = \min_{\Psi \in \mathcal{H}^1_{(N,M_S)}, \|\Psi\|_{\mathcal{L}^2}=1} \langle \Psi, H\Psi\rangle_{\mathcal{L}^2}. \quad (3.18)$$

Moreover, a normalized wave function which minimizes (3.18) corresponds to the lowest state with respect to (N, M_S) and we denote it by $\Psi_{\min}^{(N,M_S)}$.[5] Now let us assume that $E_{\min}^{(N,M_S)}$ exhibits multiplicity one and let $\{V_\kappa\}_{\kappa \in \mathbb{N}}$ be an arbitrary dense family of finite-dimensional subspaces $V_\kappa \subset \mathcal{H}^1_{(N,M_S)}$. Let further E_κ and Ψ_κ denote Galerkin approximations associated with the lowest state in the finite-dimensional subspace $V_\kappa \subset \mathcal{H}^1_{(N,M_S)}$, i.e.

$$E_\kappa = \min_{\Psi \in V_\kappa, \|\Psi\|_{\mathcal{L}^2}=1} \langle \Psi, H\Psi\rangle_{\mathcal{L}^2}, \quad \Psi_\kappa = \mathrm{argmin}_{\Psi \in V_\kappa, \|\Psi\|_{\mathcal{L}^2}=1} \langle \Psi, H\Psi\rangle_{\mathcal{L}^2}. \quad (3.19)$$

Then, $E_{\min}^{(N,M_S)} \leq E_\kappa$ for all $\kappa \in \mathbb{N}$ and thereby a relation between an estimate for the accuracy of an eigenfunction and an estimate for the approximation error of the lowest eigenvalue can be deduced, i.e. there exist $C_1, C_2 > 0$ and $\tilde{\kappa} \in \mathbb{N}$ such that the relation

$$0 \leq E_{\min}^{(N,M_S)} - E_\kappa \leq C_1 \langle \Psi_{\min}^{(N,M_S)} - \Psi_\kappa, H(\Psi_{\min}^{(N,M_S)} - \Psi_\kappa)\rangle_{\mathcal{L}^2} \leq C_2 \|\Psi_{\min}^{(N,M_S)} - \Psi_\kappa\|_{\mathcal{H}^1}^2 \quad (3.20)$$

holds for all $\kappa \geq \tilde{\kappa}$; see [161, 192]. For more details of the variational formulation of the eigenvalue problem (3.12) see also [192, 196].

3.4 Properties of the solution

In the following we briefly review important properties of the solution of the N-electron Schrödinger equation (3.9).

3.4.1 Discrete spectrum and exponential bounds

We consider the spectrum $\sigma(H)$ of an electronic Hamilton operator H for molecules as in (3.3). In particular, the operator H is semibounded, self-adjoint and its discrete

---

[4]The bilinearform $\langle \cdot, H \cdot \rangle_{\mathcal{L}^2}$ can be extended to a bounded, symmetric and coercive bilinearform on $\mathcal{H}^1$ by a shift; see e.g. [192, 196].

[5]Note that the so-called ground-state $\Psi_{\text{ground}}^{(N,M_S)}$ is associated with the minimal eigenvalue $E_{\text{ground}}^{(N,M_S)} := \min_{-N/2 \leq M_S \leq N/2} E_{\min}^{(N,M_S)}$. For a spin-independent electronic Hamiltonian operator it is usually given for $M_S = 0$ in the case of even N and for $M_S = \pm 1/2$ in the case of odd N.

spectrum $\sigma_{disc}(H)$ is defined by the set of all isolated eigenvalues of finite multiplicity. Furthermore, the essential spectrum $\sigma_{ess}(H)$ is defined as the complement of the discrete spectrum $\sigma_{ess}(H) := \sigma(H) \setminus \sigma_{disc}(H)$. Note further that in quantum mechanics the discrete spectrum corresponds to the so-called bound-states, whereas the so-called free-states correspond to the absolutely continuous spectrum. In this thesis we are only interested in the discrete spectrum and in particular in the ground-state energy $E_0 = \inf \sigma(H)$. Let us add that in quantum mechanics the so-called exited-states correspond to the eigenfunctions with eigenvalues $E \in \sigma_{disc}(H)$ and $E > E_0$.

The basis for all variational methods applied to the discrete spectrum constitutes the so-called HVZ (Hunziker, van Winter, and Zhislin) theorem [98]. From this theorem, it follows in the case of an N-electron electronic Hamilton operator H that the essential spectrum is given by $\sigma_{ess}(H) = [\Sigma, \infty)$, where the lower energy bound is equal to $\Sigma = \inf \sigma(\tilde{H}) \leq 0$. Here, $\tilde{H}$ denotes the electronic Hamilton operator which corresponds to the fixed arrangement of nuclei associated with the electronic operator H but with one electron less.[6] In this way, $\tilde{H}$ is an $(N-1)$-electron Hamilton operator and thus Σ is the so-called ionization threshold. In particular, if $N \leq \sum_{q=1}^{N_{nuc}} Z_q$ holds for a system, e.g. in the case of atoms, molecules and positive ions, then the discrete spectrum is only below the essential spectrum, i.e. $E_0 \leq E < \Sigma$ for all $E \in \sigma_{disc}(H)$, and the discrete spectrum consists of infinitely many eigenvalues [98, 163]. On the other hand it is known that in the case of $N \geq N_{nuc} + 2\sum_{q=1}^{N_{nuc}} Z_q$ the discrete spectrum is empty [124]. Furthermore, eigenfunctions which are associated with eigenvalues of the electronic Hamilton operator in the discrete spectrum are known to decay exponentially [2]. In particular, the exponential decay of a wave function Ψ is described by an $\mathcal{L}^2$ exponential bound, i.e. there is a positive function h with

$$\int_{(\mathbb{R}^3)^N} e^{h(\vec{x})} |\Psi(\vec{x})|^2 \, d\vec{x} < \infty. \tag{3.21}$$

Note that in general, an in some sense optimal bound should be anisotropic [2, 98]. To this end, Agmon expressed in his seminal work [2] an anisotropic bounded function h as a geodesic distance in terms of a certain Riemannian metric which takes different ionization thresholds into account. With the help of this so-called Agmon distance the anisotropic exponential decay of the eigenfunctions associated with eigenvalues in the discrete spectrum can be described accurately.

As an example from [2], we recall the case of an atom within the Born-Oppenheimer approximation. Here, Agmon studies in detail the $\mathcal{L}^2$-decay of the eigenfunctions of the electronic Hamiltonian H of an atom with one nucleus fixed in the origin of the coordinate system. To this end, for $I \subset \{1, \ldots, N\}$ let H_I denote the restriction of the full Hamiltonian H to the subsystem involving only the electrons associated with I and $\Lambda_I = \inf \sigma(H_I)$, $\Lambda_I = 0$ if I is empty. For any $\vec{x} \in (\mathbb{R}^D)^N \setminus \{\vec{0}\}$ let $I(\vec{x})$ denote the subset of integers $p \in \{1, \ldots, N\}$ for which $\mathbf{x}_p = \mathbf{0}$. Now, for eigenfunctions Ψ with an eigenvalue E in the discrete spectrum of H, a characterization of the type (3.21) with a positive function

$$h_{aniso} : (\mathbb{R}^3)^N \to \mathbb{R} : \vec{x} \mapsto 2(1-\varepsilon)\rho(\vec{x})$$

for any $\varepsilon > 0$ is given in [2]. Here, $\rho(\vec{x})$ is the geodesic distance from $\vec{x}$ to the origin in

---
[6]Note that the lowest energy of a system with $N = 0$ is equal to zero.

3 Electronic Schrödinger Equation

the Riemannian metric

$$d\vec{s}^2 = (\Lambda_{I(\vec{x})} - E) \sum_{i=1}^{N} 2|d\mathbf{x}_p|_2^2.$$

Note that ρ is *not isotropic*. It takes into account the amount of electrons with position $\mathbf{0}$, i.e. the number of electron-nucleus cusps, at each point $\vec{x}$.

However, there are also useful *isotropic* bounds. For example, if $\Psi^{(N,M_S)} \in \mathcal{H}^1_{(N,M_S)}$ is a weak solution of the eigenvalue problem (3.12) with the electronic Hamiltonian for molecules (3.3), and if the associated eigenvalue is below the ionization threshold $\Sigma^{(N,M_S)}$, i.e. $E^{(N,M_S)} < \Sigma^{(N,M_S)}$, then $\Psi^{(N,M_S)}$ and $\nabla\Psi^{(N,M_S)}$ decay exponentially in the $\mathcal{L}^2$-sense, i.e.

$$\int_{(\mathbb{R}^3)^N} e^{h_{iso}(\vec{x})} |\Psi^{(N,M_S)}(\vec{x})|^2 \, d\vec{x} < \infty \text{ and } \int_{(\mathbb{R}^3)^N} e^{h_{iso}(\vec{x})} |\nabla\Psi^{(N,M_S)}(\vec{x})|_2^2 \, d\vec{x} < \infty$$

with

$$h_{iso} : (\mathbb{R}^3)^N \to \mathbb{R} : \vec{x} \mapsto \sqrt{2(\Sigma' - E^{(N,M_S)})}|\vec{x}|_2$$

for $E^{(N,M_S)} < \Sigma' < \Sigma^{(N,M_S)}$; see [196] and selected references therein. For more details on the general basics of Schrödinger operators and the quantum N-body problem see the review articles [98, 163]. In particular, with respect to Coulomb systems see [125] and concerning the anisotropic exponential decay of the bound-states see [2].

3.4.2 Cusp conditions and regularity results

At first, let us consider the eigenvalue problem

$$H\Psi^{sl} = E^{sl}\Psi^{sl} \tag{3.22}$$

with the spin-independent electronic Hamilton operator (3.3) and without any side conditions due to spin. Note that the spatial electronic wave functions $\Psi^{(N,M_S)}$ according to the eigenvalue problem (3.12) also solve the eigenvalue problem (3.22). Since in the case of $D = 3$ the Coulomb potential $\frac{1}{|\mathbf{x}|_2}$ is only unbounded at $\mathbf{x} = 0$, the interaction potentials V_{ne} and V_{nn} are only singular at the set of coalescence points

$$\mathcal{C} := \left\{ \vec{x} \in (\mathbb{R}^3)^N : \left(\prod_{p=1}^{N} \prod_{q=1}^{N_{\text{nuc}}} |\mathbf{x}_p - \mathbf{R}_q|_2 \right) \left(\prod_{p=1}^{N} \prod_{p'>p}^{N} |\mathbf{x}_p - \mathbf{x}_{p'}|_2 \right) = 0 \right\}. \tag{3.23}$$

Thus, the spinless eigenfunctions

$$\Psi^{sl} : (\mathbb{R}^3)^N \to \mathbb{R}$$

are nonanalytic on $\mathcal{C}$ and analytic elsewhere $(\mathbb{R}^3)^N \setminus \mathcal{C}$.

In 1957, Kato proved that the spinless N-electron wave functions Ψ^{sl} are locally Lipschitz [104]. Moreover, in [104] he analyzed the behaviour of spinless N-electron wave functions Ψ^{sl} of an atom near coalescence points $\vec{r}^{\mathcal{C}} \in \mathcal{C}$ with exactly one singular term in the interaction potential $V_{ne} + V_{nn}$, the so-called two-particle coalescence points. By assuming that Ψ^{sl} does not vanish at the coalescence point $\vec{r}^{\mathcal{C}} \in \mathcal{C}$, i.e. $\Psi^{sl}(\vec{r}^{\mathcal{C}}) \neq 0$, he proved the so-called *cusp conditions*, i.e. conditions an eigenfunction has to obey at

3.4 Properties of the solution

a coalescence point $\vec{r}^{\mathcal{C}} \in \mathcal{C}$. For example, Kato's cusp conditions in the case of an N-electron atom of charge Z centered at the origin for coalescence points $\vec{r}^{\mathcal{C}} \in \mathcal{C}$ and an electron-nucleus cusp with $\mathbf{r}_p^{\mathcal{C}} = 0$ (w.l.o.g. let $p = 1$) reads as

$$\left.\frac{\partial \tilde{\Psi}^{sl}}{\partial r_1}\right|_{r_1=0} = -Z\Psi^{sl}\left(0, \mathbf{r}_2^{\mathcal{C}}, \ldots, \mathbf{r}_N^{\mathcal{C}}\right), \tag{3.24}$$

where $r_1 = |\mathbf{r}_1|_2$ and $\tilde{\Psi}^{sl}$ denotes the spherical average of Ψ^{sl} over an infinitesimally small sphere at $\mathbf{r}_1^{\mathcal{C}} = 0$. Concerning an electron-electron cusp for $\vec{r}_i^{\mathcal{C}} - \vec{r}_j^{\mathcal{C}} = 0$ (w.l.o.g. let $i = 1$ and $j = 2$) the cusp condition can be written in the form

$$\left.\frac{\partial \tilde{\tilde{\Psi}}^{sl}}{\partial r_{12}}\right|_{r_{12}=0} = \tfrac{1}{2}\Psi^{sl}\left(\tfrac{1}{2}(\mathbf{r}_1 + \mathbf{r}_2), \tfrac{1}{2}(\mathbf{r}_1 + \mathbf{r}_2), \mathbf{r}_3^{\mathcal{C}}, \ldots, \mathbf{r}_N^{\mathcal{C}}\right), \tag{3.25}$$

where $r_{12} = |\mathbf{r}_1 - \mathbf{r}_2|_2$ and $\tilde{\tilde{\Psi}}^{sl}$ denotes the spherical average of Ψ^{sl} over an infinitesimal small sphere about $\tfrac{1}{2}(\mathbf{r}_1 + \mathbf{r}_2)$ with constant r_{12}. Note that due to the partially antisymmetry conditions (3.12), the spatial part of an N-electron wave function $\Psi^{(N,M_S)}$ vanishes at a coalescence point of more than two electrons. Thus, besides the case of molecules and the case of coalescence points of more than two particles, Kato's cusp conditions have been particularly generalized to the case of an eigenfunction Ψ^{sl} which vanishes at the considered coalescence point $\Psi^{sl}(\vec{r}^{\mathcal{C}}) = 0$. This has been done by several authors, e.g. by Pack et al. [149], Hoffmann-Ostenhof et al. [94, 95] and Tew [174]. Let us recall now some of their main results.

In [149] for two-particle cusp conditions Pack et al. derive an extension to Kato's results by analysing an expansion in terms of real spherical harmonics of a spinless eigenfunction of (3.22) in the vicinity of any two particles' coalescence. Especially, the case of an N-electron wave function Ψ^{sl}, which may vanish at the considered two-particle coalescence point $\vec{r}^{\mathcal{C}}$, is also included. In the case of $\Psi^{sl}(\vec{r}^{\mathcal{C}}) \neq 0$ for a given two-particle electron-electron coalescence point $\vec{r}^{\mathcal{C}} \in \mathcal{C}$ the cusp condition

$$\Psi^{sl}\left(\mathbf{r}_1, \mathbf{r}_2, \mathbf{r}_3^{\mathcal{C}}, \ldots, \mathbf{r}_N^{\mathcal{C}}\right) = \Psi^{sl}\left(\tfrac{1}{2}(\mathbf{r}_1 + \mathbf{r}_2), \tfrac{1}{2}(\mathbf{r}_1 + \mathbf{r}_2), \mathbf{r}_3^{\mathcal{C}}, \ldots, \mathbf{r}_N^{\mathcal{C}}\right)\left(1 + \tfrac{1}{2}r_{12}\right) + \mathcal{O}\left(r_{12}^2\right) \tag{3.26}$$

holds, and in the case that Ψ^{sl} vanishes by first order in $\vec{r}^{\mathcal{C}}$, there follows the cusp condition

$$\Psi^{sl}\left(\mathbf{r}_1, \mathbf{r}_2, \mathbf{r}_3^{\mathcal{C}}, \ldots, \mathbf{r}_N^{\mathcal{C}}\right) = \vec{r}_{12}^T \left.\frac{\partial \Psi^{sl}}{\partial \vec{r}_{12}}\right|_{r_{12}=0}\left(1 + \tfrac{1}{4}r_{12}\right) + \mathcal{O}\left(r_{12}^3\right). \tag{3.27}$$

Note that the electron-electron cusp condition (3.26) is valid for the singlet state helium, i.e. the state which possesses minimal energy for two electrons of opposite spin, and the electron-electron cusp condition (3.26) is valid for the triplet state of helium, i.e. the state which possesses minimal energy for two electrons of same spin. In [157] for the first order derivatives of a spherically averaged wave function, Kato's cusp conditions (3.24) and (3.25) were extended to the case of the third order derivatives of a spherically averaged wave function. More recently, the results of Pack et al. were improved by Tew [174]. Here, the structure of the wave function is examined to second and higher orders in vicinity of the coalescence of any two charged particles, if these are well separated from all other particles.

The case of many-particle coalescence and the influence of partially antisymmetry due to spin is considered by Hoffmann-Ostenhof et al. in [94, 95]. In these works the authors

3 Electronic Schrödinger Equation

show that for any given coalescence point $\vec{r}^{\mathcal{C}} \in \mathcal{C}$, there exists a harmonic homogeneous polynomial $P_{M_P} \neq 0$ of degree M_P, i.e. $\Delta P_{M_P} = 0$ and $P_{M_P}(\lambda \vec{r}) = \lambda^{M_P} P_{M_P}(\vec{r})$ for $\lambda \in \mathbb{R}$, which describes the behaviour of a spinless N-electron wave function Ψ^{sl} for $|\vec{r} - \vec{r}^{\mathcal{C}}|_2 \to 0$. Then the cusp condition

$$\Psi^{sl}(\vec{r}) = P_{M_P}\left(\vec{r} - \vec{r}^{\mathcal{C}}\right) + \mathcal{O}\left(|\vec{r} - \vec{r}^{\mathcal{C}}|_2^{M_P+1}\right) \tag{3.28}$$

holds near the coalescence point $\vec{r}^{\mathcal{C}}$, and P_{M_P} can be written as

$$P_{M_P}(\vec{r}) = |\vec{r}|_2^{M_P} Y_{M_P}\left(\tfrac{1}{|\vec{r}|_2} \vec{r}\right),$$

where Y_{M_P} denotes a multi-dimensional hyperspherical harmonic. In addition, Hoffmann-Ostenhof et al. consider the minimal order of zeroes in a coalescence point in the case of a spatial N-electron wave function $\Psi^{(N,M_S)}$ with partial antisymmetry conditions according to (3.12). Note that the symmetry conditions on the wave function in (3.28) carry over to P_{M_P}. Hence, the minimal vanishing order of $\Psi^{(N,M_S)}$ in a given coalescence point $\vec{r}^{\mathcal{C}} \in \mathcal{C}$ is equal to the minimal degree of P_{M_P} in (3.28) such that P_{M_P} satisfies the partial antisymmetry conditions according to (3.12) and $P_{M_P} \not\equiv 0$. Therefore, Hoffmann-Ostenhof et al. give the minimal degree of P_{M_P} in dependence of the number of electrons N and the total spin projection M_S. In addition, they show that the minimal degree increases for large N like $\mathcal{O}(N^{\frac{4}{3}})$.

Furthermore, a theorem about the regularity of the spinless N-electron wave functions was recently shown by Fournais et al. in [55]:

Theorem 3.1 (Fournais et al. [55]). *Let*

$$\Psi^{sl} : (\mathbb{R}^3)^N \to \mathbb{R} : \vec{x} \mapsto \Psi^{sl}(\vec{x})$$

be an eigenfunction according to the spinless eigenvalue problem (3.22) and let

$$F_{(en,ee,een)} : (\mathbb{R}^3)^N \to \mathbb{R} : \vec{x} \mapsto F_{(en)}(\vec{x}) + F_{(ee)}(\vec{x}) + F_{(een)}(\vec{x})$$

with the electron-nuclei correlation term

$$F_{(en)}(\vec{x}) := \sum_{p=1}^{N} f_{(en)}(\mathbf{x}_p), \quad f_{(en)}(\mathbf{x}) := -\sum_{q=1}^{N_{nuc}} Z_q |\mathbf{x} - \mathbf{R}_k|_2,$$

the electron-electron correlation term

$$F_{(ee)}(\vec{x}) := \sum_{p=1}^{N} \sum_{p'>p}^{N} f_{(ee)}(\mathbf{x}_p, \mathbf{x}_{p'}), \quad f_{(ee)}(\mathbf{x}, \mathbf{y}) := \frac{1}{2}|\mathbf{x} - \mathbf{y}|_2$$

and the electron-electron-nuclei correlation term

$$F_{(een)}(\vec{x}) := \sum_{p=1}^{N} \sum_{p'>p}^{N} f_{(een)}(\mathbf{x}_p, \mathbf{x}_{p'}),$$

$$f_{(een)}(\vec{x}) := \frac{2-\pi}{6\pi} \sum_{q=1}^{N_{nuc}} (\mathbf{x} - \mathbf{R}_q)^T (\mathbf{y} - \mathbf{R}_q) \ln\left(|\mathbf{x} - \mathbf{R}_q|_2^2 + |\mathbf{y} - \mathbf{R}_q|_2^2\right).$$

3.4 Properties of the solution

Then the eigenfunction Ψ^{sl} can be expressed in the form

$$\Psi^{sl} = e^{F_{(en,ee,een)}} \Phi^{sl}_{(en,ee,een)}$$

with a function

$$\Phi^{sl}_{(en,ee,een)} : (\mathbb{R}^3)^N \to \mathbb{R} : \vec{\mathbf{x}} \mapsto \Phi^{sl}_{(en,ee,een)}(\vec{\mathbf{x}})$$

in the Hölder space $C^{1,1}((\mathbb{R}^3)^N)$.[7]

This representation is optimal in the following sense: There is no other function F which is dependent on N, $\vec{\mathbf{R}}$ and $\vec{Z}$ only, but not on Ψ^{sl} or E^{sl}, such that $\Psi^{sl} = F\Phi^{sl}$ with a function Φ^{sl} having more regularity than $C^{1,1}$.

Note that expressing a wave function as a product $\Psi = F\Phi$ is common in computational quantum physics and quantum chemistry, where F is a so-called Jastrow factor. Especially, in the case of an N-electron wave function F is usually assumed to be symmetric with respect to the permutation of particles and Φ is assumed to be antisymmetric (or partially antisymmetric) due to Pauli's principle. Note further that the ansatz $\Psi^{sl} = e^{F_{(en)}+F_{(ee)}} \Phi^{sl}_{(en,ee)}$, where $\Phi^{sl}_{(en,ee)} \in C^{1,\alpha}$ for $\alpha \in (0,1)$, leads to a more general formulation of Kato's cusp conditions

$$\nabla \Psi^{sl} - \Psi^{sl}(\nabla F_{(en)} + \nabla F_{(ee)}) \in C^{0,\alpha} \text{ for } 0 < \alpha < 1.$$

Moreover, by including electron-electron-nuclei coalescence by the function $F_{(een)}$ Theorem 3.1 results in

$$\nabla \Psi^{sl} - \Psi^{sl}(\nabla F_{(en)} + \nabla F_{(ee)} + \nabla F_{(een)}) \in C^{0,1}, \quad (3.29)$$

which can be viewed as a cusp condition for second order derivatives [55]. Additionally, in [55], with respect to the regularity of the wave function near the zero-set

$$\mathcal{N}(\Psi^{sl}) := \left\{ \vec{\mathbf{x}} \in (\mathbb{R}^3)^N : \Psi^{sl}(\vec{\mathbf{x}}) = 0 \right\},$$

Theorem 3.1 implies that $\nabla \Psi^{sl} : \mathcal{N}(\Psi^{sl}) \mapsto (\mathbb{R}^3)^N$ is locally Lipschitz in contrast to Ψ^{sl}, which is just locally $\mathcal{L}^\infty$ in $\mathcal{C} \setminus \mathcal{N}(\Psi^{sl})$. Further results on the regularity of the eigenfunctions of the Schrödinger operator and on the behavior of a many-electron wave function in the neighborhood of the coalescence points can also be found in [54, 56].

Recently, in [52] Flad et al. discuss the application of the best M-term approximation theory [41] within the framework of a Jastrow-type ansatz for a wave function $\Psi = F\Phi$, where the symmetric Jastrow factor J is assumed to be given in exponential form $J = e^{F_{(1,...,N)}}$ with a decomposition of the symmetric function $F_{(1,...,N)}$ into many-electron correlation functions.[8] This decomposition is a special case of the particle-wise decompositions introduced in Section 2.4 and reads as

$$F_{(1,...,N)}(\vec{\mathbf{x}}) = \sum_{p=1}^{N} f_{(1)}(\mathbf{x}_p) + \sum_{p=1}^{N} \sum_{p'>p}^{N} f_{(2)}(\mathbf{x}_p, \mathbf{x}_{p'}) + \cdots + f_{(N)}(\mathbf{x}_1, \ldots, \mathbf{x}_N).$$

---

[7] The definition of the Hölder spaces $C^{m,\alpha}$ with $m \in \mathbb{N}_0$, $\alpha \in [0,1]$ is recalled in Appendix A.1.

[8] A Jastrow factor in exponential form $J = e^{F_{(1,...,N)}}$ is commonly used in quantum mechanics since then J provides the so-called cluster property, and the so-called size-consistency is guaranteed for extended systems; see [52] and selected references therein.

3 Electronic Schrödinger Equation

In [52] the Besov regularity, which is related to the optimal convergence rate of best M-term approximations [41], of two-electron correlation functions

$$f_{(2)} : (\mathbb{R}^3)^2 \to \mathbb{R} : (\mathbf{x}, \mathbf{y}) \mapsto f_{(2)}(\mathbf{x}, \mathbf{y})$$

is analyzed for certain natural assumptions on the asymptotic behaviour of $f_{(2)}$ near electron-electron and electron-nucleus cusps. Note that the two-electron correlation functions $f_{(2)}$ typically provide the dominant contribution to the Jastrow factor J since one-electron contributions to the wave function Ψ are usually included within the antisymmetric (or partially antisymmetric) factor Φ and since contributions to the Jastrow factor J according to three-electron and higher order terms $f_{(n \geq 3)}$ can normally be assumed to be negligibly small. In particular, Flad et al. consider the best M-term approximation spaces $A_q^\alpha(\mathcal{H}^1)$ due to Nitsche [146] for certain anisotropic tensor product wavelet bases to prove that $f_{(2)} \in A_q^\alpha(\mathcal{H}^1)$ for $q > 1$ and $\alpha = \frac{1}{q} - \frac{1}{2}$. In this way, they deduce that there exists an upper bound for the approximation error in the $\mathcal{H}^1$-norm with respect to the number M of basis functions in the framework of the best M-term approximation of two-electron correlation functions. Here, an upper bound is given by

$$\inf \left\{ \left\| f_{(2)} - f_{(2)}^M \right\|_{\mathcal{H}^1} : f_{(2)}^M = \sum_{j=1}^M c_j \phi_{\mu_j} \right\} \lesssim M^{-\frac{1}{2} + \epsilon}, \quad \forall \epsilon > 0,$$

where $\{\phi_\mu : \mu \in \mathbb{N}_0\}$ is a given anisotropic wavelet basis of certain type which possesses a number of vanishing moments equal to or greater than three.

3.4.3 Decay of weak mixed derivatives

This thesis is basically motivated by the regularity results of Yserentant with respect to the eigenfunctions of the electronic Hamilton operator. Thus, we recall his mayor results from [193, 195, 196].

Let us first consider an arbitrary eigenfunction of the electronic Hamilton operator which obeys Pauli's principle.

Theorem 3.2 (Yserentant [193]). *Let $\Psi^{(N,M_S)} \in \mathcal{H}^1_{(N,M_S)}$ a weak solution of the electronic Schrödinger equation (3.16). Then $\Psi^{(N,M_S)}$ is in $\mathcal{H}^{\frac{1}{2},1}_{\mathrm{mix}}$ and even in $\mathcal{H}^{1,1}_{\mathrm{mix}}$ for an electronic wave function of totally parallel spin, i.e. $M_S \in \{+\frac{N}{2}, -\frac{N}{2}\}$.*

To be precise, Yserentant shows in [193] that an eigenfunction $\Psi^{(N,M_S)}$ associated with a spin distribution $\vec{s}^{(N,M_S)}$ has certain square integrable mixed derivatives of order up to $N_\uparrow + 1$

$$\int_{(\mathbb{R}^3)^N} (w_{\mathrm{mix},\frac{1}{2}}(\vec{\mathbf{k}}))^2 (w_{\mathrm{iso}}(\vec{\mathbf{k}}))^2 |\hat{\Psi}^{(N,M_S)}(\vec{\mathbf{k}})|^2 \, d\vec{\mathbf{k}} < \infty$$

with respect to the spatial coordinates $\mathbf{x}_1, \ldots, \mathbf{x}_{N_\uparrow}$ and certain square integrable mixed derivatives of order up to $N_\downarrow + 1$

$$\int_{(\mathbb{R}^3)^N} (w_{\mathrm{mix},-\frac{1}{2}}(\vec{\mathbf{k}}))^2 (w_{\mathrm{iso}}(\vec{\mathbf{k}}))^2 |\hat{\Psi}^{(N,M_S)}(\vec{\mathbf{k}})|^2 \, d\vec{\mathbf{k}} < \infty$$

with respect to the spatial coordinates $\mathbf{x}_{N_\uparrow+1}, \ldots, \mathbf{x}_N$.[9] Here, $w_{\mathrm{mix},s}$ is given by

$$w_{\mathrm{mix},s}(\vec{\mathbf{k}}) := \begin{cases} \sqrt{\prod_{p=1}^{N_\uparrow}(1+(\omega(\mathbf{k}_p))^2)} & \text{for } s = +\frac{1}{2}, \\ \sqrt{\prod_{p=N_\uparrow+1}^{N}(1+(\omega(\mathbf{k}_p))^2)} & \text{for } s = -\frac{1}{2} \end{cases} \quad (3.30)$$

Hence, a fully antisymmetric solution, i.e. $M_S \in \{+\frac{N}{2}, -\frac{N}{2}\}$, possesses $\mathcal{H}_{\mathrm{mix}}^{1,1}$-regularity. Furthermore, in the case of an arbitrarily chosen total spin projection $-\frac{N}{2} \leq M_S \leq \frac{N}{2}$, the inequality

$$\int_{(\mathbb{R}^3)^N} w_{\mathrm{mix}}(\vec{\mathbf{k}})(w_{\mathrm{iso}}(\vec{\mathbf{k}}))^2 |\hat{\Psi}^{(N,M_S)}(\vec{\mathbf{k}})|^2 \, d\vec{\mathbf{k}}$$
$$\leq \frac{1}{2} \sum_{s \in \{-\frac{1}{2},\frac{1}{2}\}} \int_{(\mathbb{R}^3)^N} (w_{\mathrm{mix},s}(\vec{\mathbf{k}}))^2 (w_{\mathrm{iso}}(\vec{\mathbf{k}}))^2 |\hat{\Psi}^{(N,M_S)}(\vec{\mathbf{k}})|^2 \, d\vec{\mathbf{k}}$$

holds due to the elementary relation

$$\prod_{j=1}^{N} |\tilde{\omega}(\mathbf{k}_j)| \leq \frac{1}{2} \prod_{j=1}^{N_\uparrow} |\tilde{\omega}(\mathbf{k}_j)|^2 + \frac{1}{2} \prod_{j=N_\uparrow+1}^{N} |\tilde{\omega}(\mathbf{k}_j)|^2$$

for $0 \leq N_\uparrow \leq N$ with $\tilde{\omega} : \mathbb{R}^3 \to \mathbb{R} : \mathbf{k} \mapsto \sqrt{1+|\omega(\mathbf{k})|^2}$. Thus, any partially antisymmetric wave function possesses at least $\mathcal{H}_{\mathrm{mix}}^{\frac{1}{2},1}$-regularity.

In the following we consider the approximation error with respect to general hyperbolic cross spaces $\mathcal{V}_{\mathcal{K}_K^T}$. Here, with $D = 3$, $t' = 0$ and $r = 1$, according to Theorem 3.2 and Lemma 2.1, the estimate

$$\|\Psi^{(N,M_S)} - \mathcal{P}_{\mathcal{K}_K^T}[\Psi^{(N,M_S)}]\|_{\mathcal{H}^{r'}}$$
$$\leq \begin{cases} K^{(r'-1)-t+(Tt-(r'-1))\frac{N-1}{N-T}} \|\Psi^{(N,M_S)}\|_{\mathcal{H}_{\mathrm{mix}}^{t,1}} & \text{for } T \geq \frac{r'-1}{t}, \\ K^{(r'-1)-t} \|\Psi^{(N,M_S)}\|_{\mathcal{H}_{\mathrm{mix}}^{t,1}} & \text{for } T \leq \frac{r'-1}{t} \end{cases} \quad (3.31)$$

holds for $r' < t+1$ with $t = 1$ in the fully antisymmetric case and $t = \frac{1}{2}$ in case of an arbitrary total spin projection M_S. Especially, let us discuss the case of a fully antisymmetric solution and the resulting approximation rate with respect to general hyperbolic cross spaces $\mathcal{V}_{\mathcal{K}_K^T}$ in more detail. If we measure the approximation error in the $\mathcal{H}^1$-norm, then from Lemma 2.1 we obtain with $t' = 0$, $r' = 1$ and $t = r = 1$ the approximation order $\mathcal{O}(K^{-1+T(\frac{N-1}{N-T})})$ for $T \geq 0$ and $\mathcal{O}(K^{-1})$ for $T \leq 0$. In particular, for the choice $T = 0$ a rate of $\mathcal{O}(K^{-1})$ results. In an analogous way we can argue for the partially antisymmetric case. Then, for an arbitrarily chosen total spin projection $-\frac{N}{2} \leq M_S \leq \frac{N}{2}$ we have $\mathcal{H}_{\mathrm{mix}}^{1/2,1}$-regularity at least for the associated wave

---

[9] Here, Yserentant uses $w_{\mathrm{iso}}(\vec{\mathbf{k}}) = \sqrt{1 + \sum_{p=1}^{N}(\omega(\mathbf{k}_p))^2}$,

$$w_{\mathrm{mix},s}(\vec{\mathbf{k}}) = \begin{cases} \sqrt{1 + \prod_{p=1}^{N_\uparrow}(\omega(\mathbf{k}_p))^2} & \text{for } s = +\frac{1}{2}, \\ \sqrt{1 + \prod_{p=N_\uparrow+1}^{N}(\omega(\mathbf{k}_p))^2} & \text{for } s = -\frac{1}{2}. \end{cases}$$

and $\omega(\mathbf{k}) = |\mathbf{k}|_2$ which is (up to constants) equivalent to our definitions in (2.16), (3.30) and (2.15).

3 Electronic Schrödinger Equation

function. If we measure the approximation error in the $\mathcal{H}^1$-norm, the approximation order $\mathcal{O}(K^{-1+T\frac{N-1}{N-T}})$ for $T \geq 0$ and $\mathcal{O}(K^{-\frac{1}{2}})$ for $T \leq 0$ is obtained from Lemma 2.1 with $t' = 0$, $r' = 1$ and $t = \frac{1}{2}$, $r = 1$ ($\mathcal{H}_{\text{mix}}^{\frac{1}{2},1}$-regularity). In this case, let us consider the employment of the general sparse grid discretization scheme introduced in Section 2.3.2. To this end, we additionally assume that $\hat{\Psi}^{(N,M_S)} \in \mathcal{H}_{\text{mix}}^{\hat{t},0}$, $\hat{t} > 0$. Then, with the help of Lemma 2.8, we obtain

$$\inf_{\tilde{\Psi} \in V_{L;\tilde{J}(L)}^{0;0}} \|\Psi^{(N,M_S)} - \tilde{\Psi}\|_{\mathcal{H}^1} \lesssim \left(\frac{M}{\log_2(M)^{2(N-1)}}\right)^{-\frac{1}{6(1+c)}} \|\Psi^{(N,M_S)}\|_{\mathcal{H}_{\text{mix}}^{\frac{1}{2},1,\hat{t},0}},$$

where $c = 3/(2\hat{t})$ and $M = |V_{L;\tilde{J}(L)}^{0;0}|$ denotes the number of the involved degrees of freedom. Here, we have $c \to 0$ if the Fourier transform $\hat{\Psi}^{(N,M_S)}$ is smooth, i.e. if $\Psi^{(N,M_S)}$ decays sufficiently fast (cf. Section 2.1.3). In particular, up to logarithmic terms, the convergence rate is independent of the number of electrons N and almost the same as in the two-electron case. Thus, we obtain a rate of order $-\frac{1}{6(1+c)}$ and hence due to (3.20) a rate of order $-\frac{2}{6(1+c)}$ for the minimal eigenvalue.

However, the constants involved in a norm equivalence on $\mathcal{H}_{\text{mix}}^{t,r}$ may depend on N and D. Moreover, in (3.31) also the $\mathcal{H}_{\text{mix}}^{1,1}$- and $\mathcal{H}_{\text{mix}}^{\frac{1}{2},1}$-terms may grow exponentially with the number N of electrons. This is a serious problem for any further discretization of the general hyperbolic cross spaces $\mathcal{V}_{\mathcal{K}_K^T}$. In order to compensate for this exponential growth, the parameter K has to be chosen *dependent* on N. Such a behavior can be observed in the case of a finite domain with periodic boundary conditions with Fourier bases from the results of the numerical experiments in [69] and was one reason why problems with higher numbers of electrons could not be treated.

Recently, in [195] Yserentant suggested a rescaling of the mixed Sobolev norm and the general hyperbolic cross space. To this end, a scaled analog of the $\mathcal{H}_{\text{mix}}^{1,r}$-norm, $r \in \{0,1\}$, is introduced for $s \in \{-\frac{1}{2}, +\frac{1}{2}\}$ with a scaling parameter Θ via

$$\|\Psi^{(N,M_S)}\|_{\mathcal{H}_{\text{mix},\Theta,s}^{1,r}}^2 = \int_{(\mathbb{R}^D)^N} \left(w_{\text{mix},s}\left(\frac{\vec{k}}{\Theta}\right)\right)^2 \left(\sum_{j=1}^N \left(\omega\left(\frac{k_j}{\Theta}\right)\right)^2\right)^r |\hat{\Psi}^{(N,M_S)}(\vec{k})|^2 \, d\vec{k}, \quad (3.32)$$

where $w_{\text{mix},s}$ and ω are set corresponding to (3.30) and (2.11). Furthermore, the domain

$$\mathcal{K}_{K,\Theta}^{(N,M_S)} := \left\{\vec{k} \in (\mathbb{R}^D)^N : \left(w_{\text{mix},-\frac{1}{2}}\left(\frac{\vec{k}}{\Theta}\right)\right)^2 + \left(w_{\text{mix},\frac{1}{2}}\left(\frac{\vec{k}}{\Theta}\right)\right)^2 \leq K^2\right\} \quad (3.33)$$

in Fourier space describes a Cartesian product of two scaled regular hyperbolic crosses. In the extreme case of totally parallel spin, i.e. $M_S \in \{+\frac{N}{2}, -\frac{N}{2}\}$, it degenerates into just one scaled regular hyperbolic cross. For examples of the case of two and three particles with $D = 1$ see Figure 3.1. The projection of a function f onto the scaled hyperbolic cross space $\mathcal{V}_{K,\Theta}^{(N,M_S)}$, i.e. the space of functions with vanishing Fourier transforms outside

3.4 Properties of the solution

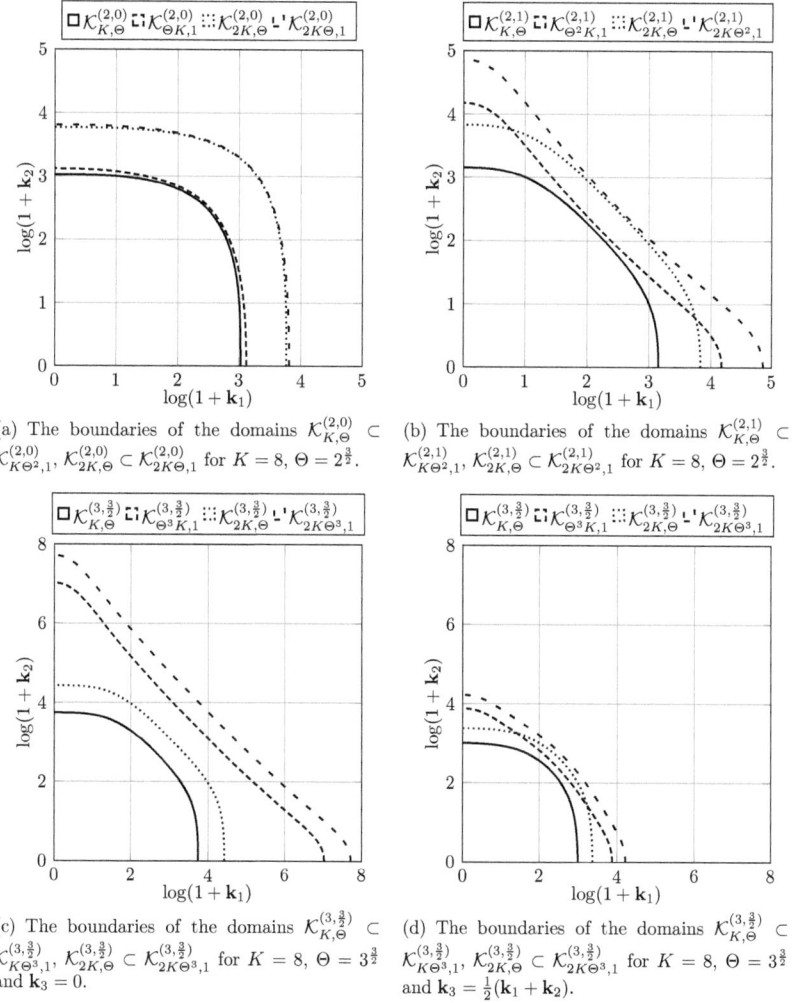

Figure 3.1: The case $D = 1$ scaled ($\Theta > 1$) and non-scaled ($\Theta = 1$): Hyperbolic cross sets $\mathcal{K}_{K,\Theta}^{(N,M_S)}$ according to (3.33) are depicted for two particles which possess opposite spin (a), for two particles which possess parallel spin (b) and for three particles which possess totally parallel spin (c)-(d).

the domain $\mathcal{K}_{K,\Theta}^{(N,M_S)}$, is given by

$$\mathcal{P}_{\mathcal{K}_{K,\Theta}^{(N,M_S)}}[f](\vec{\mathbf{x}}) = \left(\frac{1}{\sqrt{2\pi}}\right)^{3N} \int_{(\mathbb{R}^D)^N} e^{i\vec{\mathbf{x}}^T\vec{\mathbf{k}}} \chi_{\mathcal{K}_{K,\Theta}^{(N,M_S)}}(\vec{\mathbf{k}}) \hat{f}(\vec{\mathbf{k}}) \, d\vec{\mathbf{k}}$$

$$= \left(\frac{1}{\sqrt{2\pi}}\right)^{3N} \int_{\mathcal{K}_{K,\Theta}^{(N,M_S)}} e^{i\vec{\mathbf{x}}^T\vec{\mathbf{k}}} \hat{f}(\vec{\mathbf{k}}) \, d\vec{\mathbf{k}}$$

3 Electronic Schrödinger Equation

with the help of the characteristic function $\chi_{K,\Theta}^{(N,M_S)}$ on the domain $\mathcal{K}_{K,\Theta}^{(N,M_S)}$. Then, in the case of $D = 3$ with the projection $\mathcal{P}_{\mathcal{K}_{K,\Theta}^{(N,M_S)}}$, the following error estimate is shown in [195]:

Theorem 3.3 (Yserentant [195]). *Let $\Psi^{(N,M_S)} \in \mathcal{H}_{(N,M_S)}^1$ be an eigenfunction of the electronic Schrödinger equation according to (3.16) which possesses a negative eigenvalue $E < 0$. Then there exists a scaling parameter Θ such that the estimates of the approximation error measured in the $\mathcal{L}^2$-norm*

$$\|\Psi^{(N,M_S)} - \mathcal{P}_{\mathcal{K}_{K,\Theta}^{(N,M_S)}}[\Psi^{(N,M_S)}]\|_{\mathcal{L}^2} \leq \frac{2\sqrt{e}}{K}\|\Psi^{(N,M_S)}\|_{\mathcal{L}^2} \qquad (3.34)$$

and of the approximation error measured in the $\mathcal{H}^1$-seminorm

$$|\Psi^{(N,M_S)} - \mathcal{P}_{\mathcal{K}_{K,\Theta}^{(N,M_S)}}[\Psi^{(N,M_S)}]|_{\mathcal{H}^1} \leq \frac{2\sqrt{e}}{K}\Theta\|\Psi^{(N,M_S)}\|_{\mathcal{L}^2} \qquad (3.35)$$

hold.

In particular, there is a minimal scaling parameter $\Theta \lesssim \sqrt{N}\max(N,Z)$ such that the estimates (3.34) and (3.35) hold for all admissible eigenfunctions of corresponding symmetry for negative eigenvalues [195], where $Z = \sum_{q=1}^{N_{\text{nuc}}} Z_q$ is total charge of the nuclei. Especially, for an electronically neutral system we obtain $Z = N$ and thus $\Theta \lesssim N^{3/2}$. The scaling parameter Θ relates to the intrinsic length scale of the atom or molecule under consideration; see [151, 195]. Let us note that the restriction to eigenfunctions of the electronic Schrödinger Hamiltonian whose associated eigenvalues are strictly smaller than zero is not a severe issue since such an assumption holds for bound-states, i.e. states in the discrete spectrum. Also, compare Section 3.4.1 and [98, 163].

This surprising result shows that, with proper scaling in the norms and the associated choice of a scaled hyperbolic cross, it is possible to get rid of the $\|\Psi_{K,\Theta}^{(N,M_S)}\|_{\mathcal{H}_{\text{mix}}^{1,l}}$-terms on the right hand side of hyperbolic cross estimates of the type (3.31). Note that these terms may grow exponentially with N whereas $\|\Psi_{K,\Theta}^{(N,M_S)}\|_{\mathcal{L}^2} = 1$. To derive semidiscrete approximation spaces which, e.g. after scaling, overcome this problem is an important step towards any efficient discretization for problems with higher numbers of electrons. Let us remark, however, that for $\Theta \geq 1$ at least the inclusion

$$\mathcal{K}_{K,\Theta}^{(N,M_S)} \subset \mathcal{K}_{\Theta^{\max(N_\uparrow,N_\downarrow)}K,1}^{(N,M_S)}$$

holds. See Figure 3.1 for two and three particle examples in the case $D = 1$. Hence, for $\Theta \geq 1$ the scaled hyperbolic cross space $\mathcal{V}_{K,\Theta}^{(N,M_S)}$ is embedded in the non-scaled hyperbolic cross space $\mathcal{V}_{\Theta^{\max(N_\uparrow,N_\downarrow)}K,1}^{(N,M_S)}$. Here, the parameter $\Theta^{\max(N_\uparrow,N_\downarrow)}K$ of the non-scaled hyperbolic cross space grows exponentially with $\frac{N}{2} \leq \max(N_\uparrow, N_\downarrow) \leq N$.

More recently, in [196] Yserentant improved his regularity results from Theorem 3.3 towards the expansion of a bound-state in terms of tensor products of eigenfunctions $\mathcal{B} := \{\phi_\mu : \mu \in \mathbb{N}\}$ associated with increasing eigenvalues $0 < \lambda_1 \leq \lambda_2 \leq \lambda_3 \cdots$ of one particle Schrödinger operators $-\Delta + V^2$ with certain infinitely differentiable potential functions $V : \mathbb{R}^3 \to \mathbb{R}$ with $V > 0$ and $\lim_{|\mathbf{x}|_2 \to \infty} V(\mathbf{x}) = +\infty$. Here, $\mathcal{B}$ is an orthonormal basis of $\mathcal{L}^2(\mathbb{R}^3)$; see [151, 196]. Then, we can represent any $f \in \mathcal{L}^2((\mathbb{R}^3)^N)$ as

$$f(\vec{\mathbf{x}}) = \sum_{\vec{\mu} \in \mathbb{N}^N} f_{\vec{\mu}} \prod_{p=1}^{N} \phi_{\mu_p}(\mathbf{x}_p)$$

3.4 Properties of the solution

with the coefficients $f_{\vec{\mu}} := \langle \prod_{p=1}^{N} \phi_{\mu_p}, f \rangle$. In an analogous way to (3.32) and (3.33), a scaled norm is introduced by

$$\|f\|^2_{\mathcal{H}^{t,r}_{\mathrm{mix},\Theta,\mathcal{B}}} := \sum_{\vec{\mu} \in \mathbb{N}^N} \left(\left(\prod_{p=1}^{N_\uparrow} \frac{\lambda_{\mu_p}}{\Theta^2} \right)^t + \left(\prod_{p=N_\uparrow+1}^{N} \frac{\lambda_{\mu_p}}{\Theta^2} \right)^t \right) \left(\sum_{p=1}^{N} \frac{\lambda_{\mu_p}}{\Theta^2} \right)^r |f_{\vec{\mu}}|^2$$

for $t \in \{0,1\}$, $r \in \{0,1\}$, and a hyperbolic cross index set is defined by

$$\Lambda^{(N,M_S)}_{K,\Theta,\mathcal{B}} := \left\{ \vec{\mu} \in \mathbb{N}^N : \prod_{p=1}^{N_\uparrow} \frac{\lambda_{\mu_p}}{\Theta^2} + \prod_{p=N_\uparrow+1}^{N} \frac{\lambda_{\mu_p}}{\Theta^2} \leq K^2 \right\}.$$

Furthermore, the corresponding hyperbolic cross approximation space $\mathcal{V}_{\Lambda^{(N,M_S)}_{K,\Theta,\mathcal{B}}}$ is defined by the span of $\left\{ \bigotimes_{p=1}^{N} \phi_{\mu_p} : \vec{\mu} \in \Lambda^{(N,M_S)}_{M,\Theta,\mathcal{B}} \right\}$. Now, let $\Psi^{(N,M_S)} \in \mathcal{H}^1_{(N,M_S)}$ a bounded-state, i.e. an eigenfunction of the electronic Schrödinger equation according to (3.16) which possesses a negative eigenvalue $E < 0$. Then, under the assumption that $V(\mathbf{x}) \lesssim \tilde{V}(\mathbf{x})$ for all $\mathbf{x} \in \mathbb{R}^3$ with $\tilde{V}(\mathbf{x}) := \frac{1}{\tilde{\Theta}} \exp\left(\left| \frac{\mathbf{x}}{2} \right| \right)$ with a scaling parameter $\tilde{\Theta}$, there exists a scaling parameter Θ such that the norms $\|\Psi^{(N,M_S)}\|_{\mathcal{H}^{1,r}_{\mathrm{mix},\Theta,\mathcal{B}}}$, $r \in \{0,1\}$ are bounded by

$$\|\Psi^{(N,M_S)}\|^2_{\mathcal{H}^{1,r}_{\mathrm{mix},\Theta,\mathcal{B}}} \lesssim \left(\left(1 + \frac{1}{\Theta \tilde{\Theta}} \right)^2 N^2 \right)^r$$
$$\cdot \left(\sum_{s \in \{+\frac{1}{2}, -\frac{1}{2}\}} \int_{(\mathbb{R}^3)^N} (W_{\mathrm{mix},\Theta,s}(\vec{\mathbf{x}}))^2 (W_{\mathrm{iso},\Theta}(\vec{\mathbf{x}}))^2 |\Psi^{(N,M_S)}|^2 \, d\vec{\mathbf{x}} \right) < \infty$$

with

$$W_{\mathrm{iso},\Theta} := \sqrt{1 + \sum_{p=1}^{N} \left| \frac{\tilde{V}(\mathbf{x}_p)}{\Theta} \right|^2}, \quad W_{\mathrm{mix},\Theta,s} := \begin{cases} \sqrt{\prod_{p=1}^{N_\uparrow} \left(1 + \left| \frac{\tilde{V}(\mathbf{x}_p)}{\Theta} \right|^2 \right)} & \text{for } s = +\frac{1}{2}, \\ \sqrt{\prod_{p=N_\uparrow+1}^{N} \left(1 + \left| \frac{\tilde{V}(\mathbf{x}_p)}{\Theta} \right|^2 \right)} & \text{for } s = -\frac{1}{2}. \end{cases}$$

This result is shown in [196]. In addition, the following estimate with respect to the approximation error measured in the $\mathcal{H}^1$-norm is deduced:

$$\|\Psi^{(N,M_S)} - \Psi^{(N,M_S)}_{\Lambda^{(N,M_S)}_{K,\Theta,\mathcal{B}}}\|_{\mathcal{H}^1} \lesssim \|\Psi^{(N,M_S)} - \Psi^{(N,M_S)}_{\Lambda^{(N,M_S)}_{K,\Theta,\mathcal{B}}}\|_{\mathcal{H}^{0,1}_{\mathrm{mix},\Theta,\mathcal{B}}}$$
$$\leq \frac{1}{K} \|\Psi^{(N,M_S)} - \Psi^{(N,M_S)}_{\Lambda^{(N,M_S)}_{K,\Theta,\mathcal{B}}}\|_{\mathcal{H}^{1,1}_{\mathrm{mix},\Theta,\mathcal{B}}} \leq \frac{1}{K} \|\Psi^{(N,M_S)}\|_{\mathcal{H}^{1,1}_{\mathrm{mix},\Theta,\mathcal{B}}}, \tag{3.36}$$

where $\Psi^{(N,M_S)}_{\Lambda^{(N,M_S)}_{K,\Theta,\mathcal{B}}}$ denotes the best approximation of $\Psi^{(N,M_S)}$ in $\mathcal{V}_{\Lambda^{(N,M_S)}_{K,\Theta,\mathcal{B}}}$ with respect to the $\mathcal{L}^2$-norm, which is given by

$$\Psi^{(N,M_S)}_{\Lambda^{(N,M_S)}_{K,\Theta,\mathcal{B}}} = \sum_{\vec{\mu} \in \mathbb{N}^N \setminus \Lambda^{(N,M_S)}_{K,\Theta,\mathcal{B}}} \Psi^{(N,M_S)}_{\vec{\mu}} \prod_{p=1}^{N} \phi_{\mu_p}.$$

3 Electronic Schrödinger Equation

Let us note that the term $\|\Psi^{(N,M_S)}\|_{\mathcal{H}^{1,1}_{\mathrm{mix},\Theta,\mathcal{B}}}$, besides the minimal scaling parameter Θ [195, 196], does not directly depend on the decay behaviour of the Fourier transform $\hat{\Psi}^{(N,M_S)}$ in Fourier space, but only on the decay behaviour of $\Psi^{(N,M_S)}$ in real space. Thus, compared to the term $\|\Psi^{(N,M_S)}\|_{\mathcal{H}^{t,1}_{\mathrm{mix}}}$, $t \in \{\frac{1}{2},1\}$ in estimate (3.31), the term $\|\Psi^{(N,M_S)}\|_{\mathcal{H}^{1,1}_{\mathrm{mix},\Theta,\mathcal{B}}}$ in estimate (3.36) does not directly depend on the weak derivatives of the bounded-state $\Psi^{(N,M_S)}$.

Now, we suppose that $\lambda_\mu \leq C^2 \mu^{\frac{\alpha}{3}}$, where $0 < \alpha < 2$ and $C > 0$; see [196]. Then, an upper estimate for the number of degrees of freedom of the space $\mathcal{V}_{\Lambda_{K,\Theta,\mathcal{B}}^{(N,M_S)}}$ can be easily deduced, e.g. with $\bar{\Theta} = \Theta/C$, $2^L = K$ and the help of Lemma 2.2, in the form

$$\left|\Lambda_{2^L,\Theta,\mathcal{B}}^{(N,M_S)}\right| \lesssim 2^{\frac{3}{\alpha}2L}\bar{\Theta}^N \left(\frac{3}{\alpha}L + N_\uparrow \log_2(\bar{\Theta})\right)^{N_\uparrow - 1} \left(\frac{3}{\alpha}L + N_\downarrow \log_2(\bar{\Theta})\right)^{N_\downarrow - 1}$$
$$\lesssim 2^{\frac{3}{\alpha}2L} L^{N-1}. \qquad (3.37)$$

Note that we can set $\alpha = 1$ if the one-particle basis set $\mathcal{B}$ is chosen as the eigenfunctions of the harmonic oscillator, i.e. Hermite Gaussians [196].[10] Similar to Section 2.5 the subspace $\mathfrak{A}^{(N,M_S)}\left(\mathcal{V}_{\Lambda_{K,\Theta,\mathcal{B}}^{(N,M_S)}}\right) \subset \mathcal{V}_{\Lambda_{K,\Theta,\mathcal{B}}^{(N,M_S)}}$ with the partial antisymmetrizer as in (3.15), is spanned by the orthonormal set

$$\left\{\sqrt{N_\uparrow ! N_\downarrow !}\, \mathfrak{A}^{(N,M_S)} \bigotimes_{p=1}^N \phi_{\mu_p} : \vec{\mu} \in \Lambda_{K,\Theta,\mathcal{B}}^{\mathfrak{A}^{(N,M_S)}}\right\},$$

where

$$\Lambda_{K,\Theta,\mathcal{B}}^{\mathfrak{A}^{(N,M_S)}} := \left\{\vec{\mu} \in \Lambda_{K,\Theta,\mathcal{B}}^{(N,M_S)} : \mu_1 < \cdots < \mu_{N_\uparrow}, \mu_{N_\uparrow+1} < \cdots < \mu_N\right\}.$$

Compare (2.86) and (2.82). Hence, similar to Lemma 2.10 the number of degrees of freedom of the partially antisymmetric subspace is substantially reduced, i.e.

$$\left|\mathfrak{A}^{N,M_S}\left(\mathcal{V}_{\Lambda_{K,\Theta,\mathcal{B}}^{(N,M_S)}}\right)\right| \leq \frac{1}{N_\uparrow ! N_\downarrow !}\left|\mathcal{V}_{\Lambda_{K,\Theta,\mathcal{B}}^{(N,M_S)}}\right|.$$

Altogether, an upper estimate for the error measured in the $\mathcal{H}^1$-norm with respect to the involved numbers of degrees of freedom, i.e.

$$\|\Psi^{(N,M_S)} - \Psi^{(N,M_S)}_{\Lambda_{K,\Theta,\mathcal{B}}^{(N,M_S)}}\|_{\mathcal{H}^1} \lesssim \left(\frac{M}{(\ln M)^{N-1}}\right)^{-\frac{\alpha}{6}} \|\Psi^{(N,M_S)}\|_{\mathcal{H}^{1,1}_{\mathrm{mix},\Theta,\mathcal{B}}},$$

where M denotes the number of degrees of freedom $M := \left|\Lambda_{2^L,\Theta,\mathcal{B}}^{(N,M_S)}\right|$, follows from the approximation and complexity order estimates (3.36) and (3.37) analogously to Lemma 2.8. Here, up to logarithmic terms the convergence rate is independent of the number of electrons N and particularly almost the same as in the two-electron case. However, the term $\|\Psi^{(N,M_S)}\|_{\mathcal{H}^{1,1}_{\mathrm{mix},\Theta,\mathcal{B}}}$ and the constants involved in the approximation and complexity order estimates may still be exponentially dependent on the number of particles. Due to these constants and the logarithmic term, such an expansion is only practical for a moderate number of particles.

---
[10] In [196], it is also shown that one can come arbitrarily close to $\alpha = 2$, but cannot reach it.

3.5 Summary

In this chapter we gave a review on different properties of the solutions of the electronic Schrödinger equation. In particular, we recalled the regularity results of Yeserentant with respect to eigenfunctions of the electronic Hamilton operator and discussed the application of general sparse grid spaces as introduced in Section 2.3.2. Here, it turned out that the convergence rate is up to logarithmic terms independent of the number of electrons N and that it is particularly almost the same as in the two-electron case, i.e. a rate of order of $-\frac{1}{6}$ for the error of the eigenfunction measured in the $\mathcal{H}^1$-norm and thereby a rate of order of $-\frac{1}{3}$ for the minimal eigenvalue with respect to the involved number of the degrees of freedom is obtained. However, the constants involved in the approximation and complexity order estimates may still be exponentially dependent on the number of particles. Therefore, such a linear approximation scheme is only practical for a moderate number of particles.

Let us remark here that it would also be possible to treat higher numbers of particles if the wave functions would be in a weighted space equipped with finite-order weights of sufficiently low order; see Section 2.5.2. Also, an adaptive method in the framework of a nonlinear best M-term approximation approach might improve the convergence order or the involved constants substantially. However, to our knowledge there is almost no theory so far.

Let us note further that in quantum chemistry the difference of eigenvalues, e.g. ionization and binding energies, are of main interest and not the eigenvalue itself. Hence, the approximation and complexity estimates should be adapted to these cases in the future.

3 Electronic Schrödinger Equation

4 Numerical Methods

In this thesis the primarily goal is the computation of approximations to the lowest total energy (3.5) of molecular systems within the Born-Oppenheimer approximation. To this end, we apply the Galerkin discretization to approximately compute the lowest state given by the weak formulation of the electronic Schrödinger equation (3.16). Here, corresponding to (3.19), we consider Galerkin approximations in a sequence of finite-dimensional subspaces of the partially antisymmetric N-particle space. In this section we introduce our new approach for the construction of such a sequence of subspaces. Note that the application of a linear approximation scheme with the generalized sparse grid spaces introduced in Section 2.3.2 is only practical for one or two particles (cf. Section 3.4.3). In order to treat higher numbers of particles our heuristic approach is based on a particle-wise subspace splitting with finite-order weights and an h-adaptive refinement scheme.

Note that the Galerkin discretization results in a generalized linear eigenvalue problem. Hence, we first discuss the assembly of the matrices and the parallel solution of this eigenvalue problem. We further introduce an appropriate multiscale Gaussian frame to be used within the discretization scheme. Then, we introduce and discuss new finite-dimensional weighted approximation spaces. Finally, we present our new adaptive method and give all its technical details.

4.1 Galerkin discretization

Let us now discuss a Galerkin discretization of the electronic Schrödinger equation (3.12) with side conditions of partial antisymmetry. Here, according to Section 3.3 the variational formulation is used to obtain an approximation of the weak solution (3.16) in a finite-dimensional subspace $V \subset \mathcal{L}^2_{(N,M_S)} = \mathfrak{A}^{(N,M_S)}(\mathcal{L}^2((\mathbb{R}^3)^N))$. Now, let $\mathcal{B} = \{\Phi_\mu\}_{\mu \in \Lambda}$, $\Lambda = \{1, \ldots, M\}$ be a basis of the finite-dimensional subspace V. Then, the Galerkin discretization scheme results in a generalized linear eigenvalue problem in matrix form with the so-called stiffness matrix $A \in \mathbb{C}^{M \times M}$ and the so-called mass matrix $B \in \mathbb{C}^{M \times M}$. It reads as
$$Av = \tilde{E}Bv, \qquad (4.1)$$
where $v \in \mathbb{C}^M$ denotes an eigenvector and $\tilde{E} \in \mathbb{C}$ its associated eigenvalue. Here, for each pair of indices $\mu, \nu \in \Lambda$ and associated functions $\Phi_\mu, \Phi_\nu \in \mathcal{B}$ an entry in the stiffness matrix A and an entry in the mass matrix B is given by
$$(A)_{\mu,\nu} := \langle \Phi_\mu, H\Phi_\nu \rangle_{\mathcal{L}^2}, \quad (B)_{\mu,\nu} = \langle \Phi_\mu, \Phi_\nu \rangle_{\mathcal{L}^2}. \qquad (4.2)$$
An eigenvector $v = (v_1, \ldots, v_M)^T$ represents a function $\Psi^{(v)} \in V$ which reads as
$$\tilde{\Psi}^{(v)} := \sum_{\mu=1}^{M} v_\mu \Phi_\mu. \qquad (4.3)$$

4 Numerical Methods

In the case of the electronic Hamilton operator (3.3) the resulting generalized linear eigenvalue problem (4.1) is Hermitian, i.e. A is Hermitian, B is Hermitian and positive semi-definite, and hence $\tilde{E} \in \mathbb{R}$. Furthermore, as already mentioned in Section 3.2, in the case of the spin independent electronic Hamilton operator as given in (3.3) it is sufficient to consider real-valued state functions $\Psi : (\mathbb{R}^3)^N \to \mathbb{R}$ only. Hence, from now on we restrict ourselves to real-valued functions. Therefore, we can assume that $\Phi_\mu : (\mathbb{R}^3)^N \to \mathbb{R}$, $A \in \mathbb{R}^{M \times M}$, $B \in \mathbb{R}^{M \times M}$ and $v \in \mathbb{R}^M$.

For $m \leq M$ let $\Psi^{(1)}, \ldots, \Psi^{(m)}$ denote orthonormal eigenfunctions associated with the smallest m eigenvalues $E^{(1)} \leq \cdots \leq E^{(m)}$ in the discrete spectrum corresponding to (3.16).[1] Let further $v^{(1)}, \ldots, v^{(M)}$ denote B-orthonormal eigenvectors, i.e. $v^{(\mu)*} B v^{(\nu)} = \delta_{\mu,\nu}$, associated with the M eigenvalues $\tilde{E}^{(1)} \leq \cdots \leq \tilde{E}^{(M)}$ of the discrete Hermitian generalized linear eigenvalue problem (4.1). Then, as discussed in [192], the eigenvalues $\tilde{E}^{(1)} \leq \cdots \leq \tilde{E}^{(m)}$ and the functions $\tilde{\Psi}^{(v^{(1)})}, \ldots, \tilde{\Psi}^{(v^{(m)})} \in V$ given by the eigenvectors $v^{(1)}, \ldots, v^{(m)}$ with (4.3) can be seen as an approximation of the eigenvalues $E^{(1)} \leq \cdots \leq E^{(m)}$ and its associated eigenfunctions $\Psi^{(1)}, \ldots, \Psi^{(m)}$. Here, it particularly holds that $E^{(\mu)} \leq \tilde{E}^{(\mu)}$ for all $1 \leq \mu \leq M$. Note that we are primarily interested in an approximation of the lowest state, i.e. the case of $m = 1$; compare (3.19).

4.1.1 Löwdin rules

For fixed (N, M_S) we now consider the assembly of the matrices in the framework of a Galerkin approximation in a subspace of the partially antisymmetric N-particle space $\mathcal{L}^2_{(N,M_S)}$. In this thesis we compute the entries of the stiffness and the mass matrices (4.2) with the help of the so-called Löwdin rules [126] for Slater determinants. Here, an N-particle integral over a product of Slater determinant is reduced to the computation of determinants of matrices with entries put together from values of certain one- and two-particle integrals. This is explained in more detail in the following.

To this end, let us first introduce appropriate partially antisymmetric N-particle basis functions of subspaces in $\mathcal{L}^2_{(N,M_S)}$. Let $\Lambda_{(N,M_S)}$ be a finite subset of

$$\mathbb{Z}_{(N,M_S)} := \left\{ \vec{\mu} \in \mathbb{Z}^N : \mu_1 < \cdots < \mu_{N_\uparrow}, \mu_{N_\uparrow+1} < \cdots < \mu_N \right\} \tag{4.4}$$

with $|\Lambda_{(N,M_S)}| = M$ and let $\mathcal{B}_{\mathcal{L}^2} = \{\phi_\mu\}_{\mu \in \mathbb{Z}}$ be a basis of the one-particle space $\mathcal{L}^2(\mathbb{R}^3)$. Analogously to (2.82), let $V_{(N,M_S)}$ be spanned by

$$\mathcal{B}_{(N,M_S)} := \left\{ \Phi_{\vec{\mu}}^{(N,M_s)} : \vec{\mu} \in \Lambda_{(N,M_S)} \right\}, \quad \Phi_{\vec{\mu}}^{(N,M_s)} := \tilde{\mathfrak{A}}^{(N,M_S)} \bigotimes_{p=1}^{N} \phi_{\mu_p}, \tag{4.5}$$

where, similar to (2.85) and (3.15), $\tilde{\mathfrak{A}}^{(N,M_S)}$ denotes the normalized partial antisymmetrizer given by

$$\tilde{\mathfrak{A}}^{(N,M_S)} f(\vec{\mathbf{x}}) := (N_\uparrow! N_\downarrow!)^{\frac{1}{2}} \mathfrak{A}^{(N,M_S)} f(\vec{\mathbf{x}}) = (N_\uparrow! N_\downarrow!)^{-\frac{1}{2}} \sum_{P \in \mathcal{S}^{(N,M_S)}} (-1)^{|P|} f(P\vec{\mathbf{x}}). \tag{4.6}$$

Then, according to Section 2.5, $V_{(N,M_S)}$ is a finite-dimensional subspace of the partially antisymmetric N-particle space $\mathcal{L}^2_{(N,M_S)}$. Also, $\mathcal{B}_{(N,M_S)}$ is linearly independent and

---

[1] Here, we assume that for $m \geq \mu \geq 2$ all eigenvectors associated with the eigenvalues $E^{(\nu)} < E^{(\mu)}$ are in span$\{\Psi^{(1)}, \ldots, \Psi^{(\mu-1)}\}$.

4.1 Galerkin discretization

thereby builds a basis of $V_{(N,M_S)}$. Note that if $\tilde{\mathfrak{A}}$ is applied to a tensor product of one-particle functions then the resulting partially antisymmetric N-particle function can be written as a product of two Slater determinants (2.86). In this way, we obtain

$$\Phi_{\vec{\mu}}^{(N,M_s)}(\vec{\mathbf{x}}) = \tilde{\mathfrak{A}}^{(N,M_S)} \bigotimes_{p=1}^{N} \phi_{\mu_p}(\vec{\mathbf{x}})$$

$$= \frac{1}{\sqrt{N_\uparrow!}} \bigwedge_{p=1}^{N_\uparrow} \phi_{\mu_p} \otimes \frac{1}{\sqrt{N_\downarrow!}} \bigwedge_{p=N_\uparrow+1}^{N} \phi_{\mu_p}(\vec{\mathbf{x}}) \qquad (4.7)$$

$$= \frac{1}{\sqrt{N_\uparrow!}} \begin{vmatrix} \phi_1(\mathbf{x}_1) & \cdots & \phi_{N_\uparrow}(\mathbf{x}_1) \\ \vdots & \ddots & \vdots \\ \phi_1(\mathbf{x}_{N_\uparrow}) & \cdots & \phi_{N_\uparrow}(\mathbf{x}_{N_\uparrow}) \end{vmatrix} \frac{1}{\sqrt{N_\downarrow!}} \begin{vmatrix} \phi_1(\mathbf{x}_{N_\uparrow+1}) & \cdots & \phi_N(\mathbf{x}_{N_\uparrow+1}) \\ \vdots & \ddots & \vdots \\ \phi_1(\mathbf{x}_N) & \cdots & \phi_N(\mathbf{x}_N) \end{vmatrix}$$

for the basis functions, which can also be written in the form

$$\Phi_{\vec{\mu}}^{(N,M_s)}(\vec{\mathbf{x}}) = \frac{1}{\sqrt{N_\uparrow! N_\downarrow!}} \begin{vmatrix} \phi_1(\mathbf{x}_1) & \cdots & \phi_{N_\uparrow}(\mathbf{x}_1) & 0 & \cdots & 0 \\ \vdots & \ddots & \vdots & \vdots & \ddots & \vdots \\ \phi_1(\mathbf{x}_{N_\uparrow}) & \cdots & \phi_{N_\uparrow}(\mathbf{x}_{N_\uparrow}) & 0 & \cdots & 0 \\ 0 & \cdots & 0 & \phi_1(\mathbf{x}_{N_\uparrow+1}) & \cdots & \phi_N(\mathbf{x}_{N_\uparrow+1}) \\ \vdots & \ddots & \vdots & \vdots & \ddots & \vdots \\ 0 & \cdots & 0 & \phi_1(\mathbf{x}_N) & \cdots & \phi_N(\mathbf{x}_N) \end{vmatrix}. \qquad (4.8)$$

Let us now discuss the application of the Löwdin rules to compute the entries of the system matrices associated with a set of basis functions, which is given with the help of products of Slater determinants as in (4.7) or (4.8). To this end, for $k \leq N$ let o_k denote a Hermitian k-particle operator, which applies directly to a k-particle function and is symmetric with respect to permutations of particles. Now, we introduce a Hermitian N-particle operator

$$O_k := \sum_{1 \leq p_1 < \cdots < p_k \leq N} o_k(p_1, \ldots, p_k),$$

where $o_k(p_1, \ldots, p_k)$ denotes the k-particle operator which, albeit applied to an N-particle wave function, only acts on its p_1-th, ..., p_k-th particle components. Note that O_k is then also symmetric with respect to permutations of particles. For a vector of one-particle functions $\vec{\psi} = (\psi_1, \ldots, \psi_N)^T$, let $\tilde{\mathfrak{A}}^{(N,M_s)}[\vec{\psi}]$ denote the partially antisymmetric N-particle function $\tilde{\mathfrak{A}}^{(N,M_S)} \bigotimes_{p=1}^{N} \psi_p$. Now, let us consider the computation of $\langle \tilde{\Psi}, \Psi \rangle$ and $\langle \tilde{\Psi}, O_k \Psi \rangle$ for fixed partially antisymmetric N-particle functions $\tilde{\Psi} := \tilde{\mathfrak{A}}^{(N,M_s)}[\vec{\tilde{\psi}}]$, $\Psi := \tilde{\mathfrak{A}}^{(N,M_s)}[\vec{\psi}]$. Here, to shorten notation we introduce the function

$$s : \mathcal{N} \to \{+\tfrac{1}{2}, -\tfrac{1}{2}\} : p \mapsto \begin{cases} +\tfrac{1}{2} & \text{for } p \in \mathcal{N}_\uparrow, \\ -\tfrac{1}{2} & \text{for } p \in \mathcal{N}_\downarrow, \end{cases}$$

where $\mathcal{N}_\uparrow := \{1, \ldots, N_\uparrow\}$ and $\mathcal{N}_\downarrow := \{N_\uparrow+1, \ldots, N\}$. We further introduce two different matrices for the fixed $\Psi, \tilde{\Psi}$: First, an overlap matrix $\mathbf{S} \in \mathbb{R}^{N \times N}$ given by

$$(\mathbf{S})_{p,q} := \left\langle \tilde{\psi}_p, \psi_q \right\rangle_{L^2(\mathbb{R}^3)} \delta_{s(p),s(q)} \qquad (4.9)$$

and second, a k-particle operator matrix $\mathbf{O}_k \in \mathbb{R}^{\binom{N}{k} \times \binom{N}{k}}$ given by[2]

$$(\mathbf{O}_k)_{u,v} := (-1)^{|P|} \sum_{P \in \mathcal{S}_k} \mathbf{o}_k(p_1, \ldots, p_k, q_{P(1)}, \ldots, q_{P(k)}) \tag{4.10}$$

with

$$\mathbf{o}_k(p_1, \ldots, p_k, q_1, \ldots, q_k) := \left\langle \bigotimes_{j=1}^{k} \tilde{\psi}_{p_j}, \, o_k \bigotimes_{j=1}^{k} \psi_{q_j} \right\rangle_{\mathcal{L}^2((\mathbb{R}^3)^k)} \prod_{j=1}^{k} \delta_{s(p_j), s(q_j)} \tag{4.11}$$

for $u = \{p_1, \ldots, p_k\} \subset \mathcal{N}$, $p_1 < \cdots < p_k$ and $v = \{q_1, \ldots, q_k\} \subset \mathcal{N}$, $q_1 < \cdots < q_k$. Note that in [126], with the help of (3.17) and the general Laplace expansion[3], it is shown that a matrix element associated with the operator O_k with respect to two Slater determinants can be written in terms of determinants of matrices with entries put together from values of k-particle integrals. In the same way we obtain the relation

$$\left\langle \tilde{\Psi}, \, O_k \Psi_{\vec{v}} \right\rangle_{\mathcal{L}^2((\mathbb{R}^3)^N)} = \operatorname{tr}\left(\mathbf{O}_k \operatorname{adj}^{[k]}(\mathbf{S}) \right) \tag{4.12}$$

for fixed $\tilde{\Psi} = \tilde{\mathfrak{A}}^{(N,M_s)}[\vec{\tilde{\psi}}]$, $\Psi_{\vec{v}} = \mathfrak{A}^{(N,M_s)}[\vec{\psi}]$, where $\operatorname{tr}(M)$ denotes the trace of a square matrix M and $\operatorname{adj}^{[k]}(M)$ denotes the k-th order adjungate of a matrix $M \in \mathbb{R}^{N \times N}$. In particular $\operatorname{adj}^{[k]}(M)$ is in $\mathbb{R}^{\binom{N}{k} \times \binom{N}{k}}$ and for $u = \{p_1, \ldots, p_k\} \subset \mathcal{N}$, $p_1 < \cdots < p_k$ and $v = \{q_1, \ldots, q_k\} \subset \mathcal{N}$, $q_1 < \cdots < q_k$ it holds the identity

$$(\operatorname{adj}^{[k]}(M))_{u,v} = (-1)^{\sum_{q \in v} q + \sum_{p \in u} p} M^{(v,u)},$$

where $M^{(u,v)}$ denotes the (u,v) minor of M. The (u,v) minor of M is defined as the determinant of the submatrix formed by removing rows and columns from M associated with indices in u and v, respectively. In particular, for $|u| = |v| = k$, $M^{(u,v)}$ is then a so-called $(N-k)$-th order minor. Note that a k-th order cofactor is a signed $(N-k)$-th order minor, i.e. $(-1)^{\sum_{q \in v} q + \sum_{p \in u} p} M^{(u,v)}$, and a so-called k-th order compound matrix $M^{[k]}$ is formed by all minors of order $(N-k)$ of a matrix M, i.e. $(M^{[k]})_{u,v} = M^{(u,v)}$.

We can write the electronic Hamilton operator (3.3) as

$$H = O_1 + O_2,$$

where we set

$$o_1 = -\frac{1}{2}\Delta_\mathbf{x} - \sum_{j=1}^{N_{\text{nuc}}} \frac{Z_j}{|\mathbf{x} - \mathbf{R}_j|},$$

for the one-particle operator and

$$o_2 = \frac{1}{|\mathbf{x} - \mathbf{y}|}$$

for the two-particle operator. Where we clearly see that $o_2(p,q) = o_2(q,p)$. Thus, concerning the computation of the entries (4.2) for $\Phi_{\vec{\mu}}^{(N,M_s)}, \Phi_{\vec{v}}^{(N,M_s)} \in \mathcal{B}_{(N,M_s)}$, we focus

---

[2] Note that we label an entry of a $\binom{N}{k} \times \binom{N}{k}$ matrix by a pair of sets. Here, one may associate a set $u = \{p_1, \ldots, p_k\} \subset \mathcal{N}$, $p_1 < \cdots < p_k$ with the index $1 \leq \sum_{j=1}^{k} \binom{p_k}{k} \leq \binom{N}{k}$.
[3] A general Laplace expansion of a determinant of an $N \times N$ matrix M to a set $u \subset \mathcal{N}$, $|u| = k$ is given by $|M| = \sum_{v \subset \mathcal{N}, |v|=k} (-1)^{\sum_{p \in u} p + \sum_{q \in v} q} M^{(\mathcal{N} \setminus u, \mathcal{N} \setminus v)} M^{(u,v)}$, where $M^{(u,v)}$ denotes the (u,v) minor of M.

on the cases of order $k \leq 2$. We set $\tilde{\Psi} = \Phi_{\vec{\mu}}^{(N,M_S)}$, $\Psi = \Phi_{\vec{\nu}}^{(N,M_S)}$ and with the help of (4.12) we particularly obtain the relations

$$\left\langle \Phi_{\vec{\mu}}^{(N,M_S)}, \Phi_{\vec{\nu}}^{(N,M_S)} \right\rangle = |\mathbf{S}|, \tag{4.13}$$

$$\left\langle \Phi_{\vec{\mu}}^{(N,M_S)}, O_1 \Phi_{\vec{\nu}}^{(N,M_S)} \right\rangle = \operatorname{tr}\left(\mathbf{O}_1 \operatorname{adj}(\mathbf{S})\right), \tag{4.14}$$

$$\left\langle \Phi_{\vec{\mu}}^{(N,M_S)}, O_2 \Phi_{\vec{\nu}}^{(N,M_S)} \right\rangle = \operatorname{tr}\left(\mathbf{O}_2 \operatorname{adj}^{[2]}(\mathbf{S})\right), \tag{4.15}$$

where $\mathbf{S} \in \mathbb{R}^{N \times N}$ as in (4.9) and $\mathbf{O}_1 \in \mathbb{R}^{N \times N}$, $\mathbf{O}_2 \in \mathbb{R}^{\binom{N}{2} \times \binom{N}{2}}$ as in (4.10). The complexity of the computation of an entry of the mass matrix B according to (4.13) is of order $\mathcal{O}(N^3)$, since in the worst case it is equal to the complexity of the computation of a determinant of a full $N \times N$ matrix. The complexity of the computation an entry of the stiffness matrix A is dominated by the computation of the term associated with the two-particle operator. Here, a direct application of (4.15), i.e. the computation of the $\binom{N}{2}^2$ entries of $\mathbf{O}_2$ and for each entry a computation of a determinant of a $(N-2) \times (N-2)$ submatrix of $\mathbf{S}$, results in a complexity order of $\mathcal{O}(N^7)$. Note that for the computation of the term associated with the one-particle operator, the direct application of (4.14) yields a complexity order of $\mathcal{O}(N^5)$.

However, in [181], based on the work [155], a factorization of the form[4] $LSU = D$ is used for the efficient computations of cofactors. Here, L denotes an $N \times N$ matrix with $|L| = 1$, U denotes a $N \times N$ matrix with $|U| = 1$ and D denotes a diagonal $N \times N$ matrix. This, results in an overall complexity of order $\mathcal{O}(N^4)$, i.e. $\mathcal{O}(N^3)$ for (4.14) and $\mathcal{O}(N^4)$ for (4.15). Here, in the case of non-singular $\mathbf{S}$, for example, the relations $\operatorname{adj}(\mathbf{S}) = U \operatorname{adj}(D) L$ and $\operatorname{adj}^{[2]}(\mathbf{S}) = U^{[2]} \operatorname{adj}^{[2]}(D) L^{[2]}$ are utilized.

Note further that for a pair of partially antisymmetric basis functions $\tilde{\Psi} = \tilde{\mathfrak{A}}^{(N,M_s)}[\vec{\tilde{\psi}}]$ and $\Psi = \tilde{\mathfrak{A}}^{(N,M_s)}[\vec{\psi}]$, there may orthogonality be present for a subset of pairs of the involved one-particle functions, i.e. $\langle \tilde{\psi}_p, \psi_q \rangle = 0$ for $(p,q) \in \Lambda_\mathcal{O} \subset \mathcal{N}^2$. Then, the overlap matrix can be brought to block diagonal form and the block structure can be exploited to speed up the computation of the elements of the stiffness and the mass matrix [44]. Let us remark that in the special case where the one-particle functions are taken form a biorthonormal set, the overlap matrix $\mathbf{S}$ is just a permutation of a diagonal matrix D with $(D)_{p,p} \in \{0,1\}$ and hence the Löwdin rules reduce to the well-known Slater-Condon rules [35, 166]. Here, in the worst case the overall computational complexity is of order $\mathcal{O}(N^2)$.

Hence, a further possibility to gain an efficient computation of the matrix entries of the system matrices associated with a pair of partially antisymmetric N-particle functions $\tilde{\Psi} = \tilde{\mathfrak{A}}^{(N,M_s)}[\vec{\tilde{\psi}}]$ and $\Psi = \tilde{\mathfrak{A}}^{(N,M_s)}[\vec{\psi}]$ is to biorthogonalize [152] or biorthonormalize [15] the involved one-particle functions. To this end, a singular value decomposition of the overlap matrix can be employed, i.e. $D = U^* \mathbf{S} V$, where D denotes a diagonal matrix and U, V denote unitary matrices. Then, the computation of the entries of the system matrices with respect to the transformed partially antisymmetric N-particle functions $\tilde{\Psi} = \tilde{\mathfrak{A}}^{(N,M_s)}[U\vec{\tilde{\psi}}]$ and $\Psi = \tilde{\mathfrak{A}}^{(N,M_s)}[V\vec{\psi}]$ can be performed in an analogous way to the Slater-Condon rules [152]. Note however that then the two-particle integrals need to be

---

[4]In this framework a factorization $LSU = D$ is typically used which is constructed by biorthogonalization [155, 181]. Here, L is a lower triangular matrix with unit diagonal, U is an upper triangular matrix with unit diagonal and D is a diagonal matrix.

transformed as well, which again may result in an overall computational complexity of order $\mathcal{O}(N^4)$; see e.g. [44, 152, 155, 181].

Nevertheless, since we deal in this thesis with a moderate number of particles only, and the involved matrices are not sparse, we employ the Löwdin rules, i.e. relations (4.13), (4.14) and (4.15), directly. Here, we just make use of the known structure of the zero elements of the matrices due to partial antisymmetry; compare (4.9) and (4.11).

4.1.2 Parallel eigenvalue solvers

For the computation of the smallest m eigenvalues of the resulting discrete general linear eigenvalue problem (4.1), we invoke the *scalable library for eigenvalue problem computations* (SLEPc) [91]. The parallel software package SLEPc is an extension of the parallel software package PETSc [8–10], which supports the MPI standard [78] for message-passing communication. In this way, we are able to use only one interface for the application of different eigenvalue solvers together with several spectral transformations and various preconditioners. For example, SLEPc provides an interface to the well-known sequential library LAPACK [4] and to parallelized variants of the Krylov-Schur method [89, 168]. It further provides an interface to the well-known ARPACK/PARPACK [122, 131] library, which is a parallel implementation of a variant of the implicitly restarted Arnoldi method, and it particularly provides a wrapper to the software package BLOPEX, which is an implementation of the parallelized *locally optimal block preconditioned conjugate gradient* method (LOBPCG) [109]. Note that the SLEPc[5] wrapper for BLOPEX makes the LOBPCG method only available in the case of the standard eigenvalue problem ($Av = Ev$) with a real symmetric stiffness matrix ($A^* = A^T = A$). Hence, to solve the generalized linear eigenvalue problem (4.1), we improved the implementation of the BLOPEX interface in SLEPc to also deal with eigenvalue problems $Av = EBv$ with a real symmetric stiffness matrix A and a real symmetric positive definite mass matrix B.

In this thesis we mainly apply the LOBPCG method with a symmetric preconditioner $P \approx (A - \lambda B)^{-1}$ with λ chosen such that P is positive definite. Here, we perform a Cholesky decomposition of $(A - \lambda B)$ to apply the preconditioner P in an efficient way. Note that we have to treat full matrices within this thesis. Hence, we use the software package PLAPACK [180], which includes a parallel implementation of the Cholesky decomposition for full matrices.

Besides the LOBPCG method, we also employ the Krylov-Schur method as implemented in SLEPc and the Arnoldi method as implemented in the ARPACK/PARPACK library for benchmarking reasons. Here, we particularly use a spectral transformation by a shift-and-invert operator. In this way, the largest m eigenvalues of the symmetric matrix $(A - \lambda B)^{-1}B$ have to be determined and then transformed back to obtain the smallest m eigenvalues of the generalized linear eigenvalue problem $Av = EBv$ [88]. Here, we again assume that λ is chosen such that $(A - \lambda B)$ is positive definite. To apply the symmetric operator $(A - \lambda B)^{-1}$, we again employ the Cholesky decomposition as implemented in PLAPACK.

Note that we perform the assembly of the system matrices A and B in parallel in a

---

[5] Note that we use in this work the release version 2.3.3 of SLEPc, which is based on the release version 2.3.3 of PETSc.

straightforward way. Note finally that in the case of dealing with sparse system matrices A and B, one may invoke a parallelized incomplete Cholesky decomposition, e.g. as implemented in the parallel software library BlockSolve95 [102]. For a further discussion on algebraic eigenvalue problems see, for example, the practical guide [7] and the survey on software packages [90].

4.2 Multiscale Gaussian frame

Motivated by the regularity results by Yserentant [193] a hyperbolic cross approach to treat the electronic Schrödinger equation in a periodic setting using the Fourier basis is applied in [69]. Here, it turns out that the need of a global refinement for the resolution of the electron-nuclei and the electron-electron cusps leads to huge constants, which limits this approach to just one particle in practice. Note that wavelet-type frames allow for local resolution.[6] Hence, in [67] a hyperbolic cross approach similar to the one introduced in Section 2.3 based on the Meyer-wavelet family (cf. Appendix B.2) is studied in practice. Here, it turns out that, although the wavelet system allows to resolve the cusps locally, the sub-exponential decay of the basis functions in real space and the expensive numerical integration of the one-particle operator integrals and in particular of the two-particle operator integrals leads to impractical huge constants.

Therefore, in this thesis we introduce a wavelet-like frame using Gaussians, which exhibits exponential decay in real space as well as in Fourier space and in particular, allows for local adaptivity and the computation of all inner products by analytic formulae. Note that from the Balian-Low theorem there follows that no orthonormal frame with exponential decay in real space and also in Fourier space exists; see e.g. [87].

4.2.1 Gaussians in multiresolution analysis

In the following we shortly review the application of Gaussians in the framework of multiresolution analysis approaches.

To this end, let us first introduce the $\mathcal{L}^2$-normalized one-dimensional Gaussian for a given deviation parameter $\sigma > 0$ by[7]

$$g_\sigma : \mathbb{R} \to \mathbb{R} : x \mapsto \frac{1}{\sqrt{\sigma\sqrt{\pi}}} e^{-\frac{1}{2\sigma^2}x^2}, \qquad (4.16)$$

where its Fourier transform reads as

$$\hat{g}_\sigma : \mathbb{R} \to \mathbb{R} : k \mapsto \frac{1}{\sqrt{\sigma^{-1}\sqrt{\pi}}} e^{-\frac{1}{2}\sigma^2 k^2}.$$

Furthermore, for a deviation parameter $\sigma > 0$ we introduce an isotropic Gaussian in the

---

[6]For a further reading on multiresolution analysis, wavelet-type frames and wavelets see Appendix B and the monographs [32, 37, 87, 130, 139], for example.

[7]Note that the $\mathcal{L}^1$-normalized one-dimensional Gaussian $\frac{1}{\sigma\sqrt{2\pi}} e^{-\frac{1}{2\sigma^2}(x-\mu)^2}$ is the probability normal distribution of a normally distributed random variable with expected value μ, variance σ^2 and so standard deviation σ.

one-particle space $\mathcal{L}^2(\mathbb{R}^D)$ in the form

$$G_\sigma : \mathbb{R}^D \to \mathbb{R} : \mathbf{x} \mapsto \left(\bigotimes_{d=1}^{D} g_\sigma\right)(\mathbf{x}), \tag{4.17}$$

which is particularly rotationally symmetric, i.e. $G_\sigma(\mathbf{x}) = G_\sigma(\mathbf{y})$ for all $\mathbf{x}, \mathbf{y} \in \mathbb{R}^D$ with $|\mathbf{x}|_2 = |\mathbf{y}|_2$. Note that, using the tensor product structure, the rotational symmetry and the Fourier transform of a one-particle isotropic Gaussian, analytic formulae can be derived for all one- and two-particle integrals involved in the Löwdin rules for products of Slater determinants of one-particle isotropic Gaussians with different deviation parameters. For details see Appendix C.1.

Furthermore, let us recall some definitions and well-known properties of frames (cf. [32, 37, 87, 130, 139]). A family of functions $\{\phi_\mu\}_{\mu \in \Lambda}$, $\Lambda \subset \mathbb{Z}$ in a separable Hilbert space V is a frame, if there exist frame bounds $C_1, C_2 > 0$ such that for any $f \in V$ the relation

$$C_1 \|f\|_V^2 \leq \sum_{\mu \in \Lambda} |\langle \phi_\mu, f \rangle_V|^2 \leq C_2 \|f\|_V^2 \tag{4.18}$$

holds. In particular, the so-called frame operator $\mathfrak{F}[f] := \sum_{\mu \in \Lambda} \langle \phi_\mu, f \rangle_V \phi_\mu$ is then invertible. Here, the family of functions $\{\check{\phi}_\mu := \mathfrak{F}^{-1}\phi_\mu\}_{\mu \in \Lambda}$ is called the dual-frame and for every $f \in V$ it holds the reconstruction formulae

$$f = \sum_{\mu \in \Lambda} \langle \check{\phi}_\mu, f \rangle_V \phi_\mu = \sum_{\mu \in \Lambda} \langle \phi_\mu, f \rangle_V \check{\phi}_\mu.$$

A frame with frame constants $C_1 = C_2$ is called a tight frame. Note that a tight frame with frame constants $C_1 = C_2 = 1$ is in particular an orthonormal basis.

Now, let us remark that multiresolution analysis and wavelet-type decompositions are based on a family of shift-invariant spaces [42, 116]. Here, for a so-called generating function $\varphi \in \mathcal{L}^2(\mathbb{R}^D)$ the family $\{V^h(\varphi)\}_{h>0}$ is a so-called ladder of principal shift-invariant spaces generated by φ, where $V^h(\varphi)$ is defined as the $\mathcal{L}^2$-closure of span $\{\varphi(\cdot/h - \mathbf{j}) : \mathbf{j} \in \mathbb{Z}^D\}$. Let further φ be a sufficiently smooth and rapidly decaying function which satisfies the so-called moment condition: a generating function φ satisfies the moment condition of order $m \in \mathbb{N}_0$ if for $c > 0$

$$\int_{\mathbb{R}^D} \varphi(\mathbf{x})\,d\mathbf{x} = c \quad \text{and} \quad \int_{\mathbb{R}^D} \mathbf{x}^\mathbf{a} \varphi(\mathbf{x})\,d\mathbf{x} = 0, \forall \mathbf{a} \in \mathbb{N}_0^D, 1 \leq |\mathbf{a}|_1 < m, \tag{4.19}$$

which is particularly equivalent to $\hat{\varphi}(\mathbf{0}) = c$ and $D^\alpha \hat{\varphi}(\mathbf{0}) = 0, \forall \mathbf{a} \in \mathbb{N}_0^D, 1 \leq |\mathbf{a}|_1 < m$. Then, concerning the approximation properties of the ladder $\{V^h(\varphi)\}_{h>0}$, approximation rates of order m can be guaranteed up to some saturation error; see e.g. [135]. For example, it follows for $\varphi_\sigma := \varphi(\cdot/\sigma)$ that for any $\epsilon > 0$ there exists $\sigma' > 0$ such that for all $\sigma > \sigma'$ the estimate

$$\inf_{f_h \in V^h(\varphi_\sigma)} \|f - f_h\|_{\mathcal{L}^2} \lesssim ((\sigma h)^m + \epsilon) \|f\|_{\mathcal{H}^m}$$

holds for any $f \in \mathcal{H}^m(\mathbb{R}^D)$, $m > D/2$. In particular, for ϵ sufficiently small numerical computations behave like a converging approximation process within the framework of the approximation in a so-called stationary ladder $\{V^h(\varphi_\sigma)\}_{h>0}$. Moreover, in order to derive convergence, one considers a so-called nonstationary ladder $\{V^h(\varphi_{\sigma(h)})\}_{h>0}$. Here, the dilation parameter σ should be chosen dependent on h, i.e. $\sigma = \sigma(h)$, such that the saturation error is eliminated; see e.g. [135]. For a further discussion on Sobolev spaces and shift-invariant subspaces see e.g. [39, 97, 135]. For a Gaussian chosen as generating function, see especially [22]. For the stationary case, see particularly [133] and for the nonstationary case, see [100].

In order to obtain a multiresolution analysis generated by a family of one-particle functions $\{\varphi_{\sigma(l)}\}_{l \in \mathbb{Z}}$, we consider the family $\{V_l(\varphi_{\sigma(l)})\}_{l \in \mathbb{Z}}$ of shift-invariant spaces $V_l(\varphi_{\sigma(l)}) := V^{2^{-l}}(\varphi_\sigma)$. Here, the usual assumptions are that $\mathrm{clos}_{\mathcal{L}^2}(\bigcup_{l \in \mathbb{Z}} V_l) = \mathcal{L}^2(\mathbb{R}^D)$, $\bigcap_{l \in \mathbb{Z}} V_l = \{0\}$ and that the so-called refinement relation $V_l \subset V_{l+1}$ holds; compare also Appendix B.1. Then, we define a sequence of so-called detail or wavelet spaces $W_l \subset V_{l+1}$, $l \in \mathbb{Z}$ by direct sum decompositions $V_l \oplus W_l = V_{l+1}$ for all $l \in \mathbb{Z}$.[8] In this way, for example we obtain the direct sum decomposition $V_L = V_{L_0} \oplus \bigotimes_{l=L_0}^{L} W_l$ for $L_0 < L \in \mathbb{Z}$. For a further reading on a nonstationary multiresolution analysis using a Gaussian and on families of non-refinable shift-invariant spaces, see [33] and [40], respectively.

Now, if $\sigma(l)$ is held fixed for all $l \in \mathbb{Z}$, then the corresponding multiresolution analysis is called stationary, i.e. if the family $\{V_l(\varphi_\sigma)\}_{l \in \mathbb{Z}}$ is obtained by dilating the same space $V_0 = V^1(\varphi_\sigma)$ for a fixed $\sigma > 0$ (cf. Appendix B.1). In this case, the refinement relations reduce to the two-scale refinement relation $V_0 \subset V_1$ and hence, the generating one-particle function φ_σ has to satisfy the two-scale refinement equation

$$\varphi_\sigma(\mathbf{x}) = \sum_{\mathbf{j} \in \mathbb{Z}^D} c_{\mathbf{j}} \varphi_\sigma(2\mathbf{x} - \mathbf{j}).$$

Note that the generating function φ_σ is then also called scaling function. In particular, the Gaussian does not fulfill the two-scale refinement equation and thereby, is not a scaling function [87, 130]. However, an isotropic Gaussian G_σ satisfies the two-scale refinement equation up to an error, which can be made smaller than any $\epsilon > 0$ by choosing σ sufficiently large. In [134] this idea is employed to introduce an approximate stationary multiresolution analysis using Gaussians and to construct approximate wavelets. Such an approximate wavelet can be defined in terms of anisotropic Gaussians with complex-valued arguments, that is in terms of type $e^{-(A(\mathbf{x}-\mathbf{z}))^T(\mathbf{x}-\mathbf{z})}$ with an appropriate positive-definite matrix $A \in \mathbb{C}^{D \times D}$ and a vector $\mathbf{z} \in \mathbb{C}^D$. For a further reading on approximate wavelet decomposition, see [135].

Furthermore, a method for the construction of wavelet-like bases and frames with the help the well-known small perturbation principle is proposed in [118, 119]. Here, the idea is to approximate every frame element ϕ_μ of a given frame $\{\phi_\mu\}_{\mu \in \Lambda}$ for a space V by an appropriate function $\tilde{\phi}_\mu$ such that the new system $\{\tilde{\phi}_\mu\}_{\mu \in \Lambda}$ also builds a frame for the space V. In particular, the new functions $\tilde{\phi}_\mu$ can be constructed by linear combinations of a fixed small number of shifts and dilates of a sufficiently smooth and rapidly decaying generating function φ like, for example, the Gaussian function.

Such new wavelet-like systems have been in detail discussed for the case of isotropic homogeneous Besov and Triebel-Lizorkin spaces [178] in [118, 119]. Note that an obvious modification of this method produces frames for inhomogeneous Triebel-Lizorkin and Besov spaces as well. Note further that these spaces also include the isotropic

---

[8]Note that it is common to use the orthogonal direct sum decomposition here, i.e. $V_l \oplus W_l = V_{l+1}$ with $V_l \perp W_l$.

4 Numerical Methods

Sobolev spaces $\mathcal{H}^r$; compare Appendix A.2. In addition, in [119] the boundedness and invertibility of the frame operator $\tilde{\mathfrak{F}}[f] := \sum_{\mu \in \Lambda} \langle \tilde{\phi}_\mu, f \rangle \tilde{\phi}_\nu$ on Triebel-Lizorkin and Besov spaces is studied. Furthermore, necessary and sufficient conditions which determine if the membership of an $f \in \mathcal{S}'$ of a certain Triebel-Lizorkin or Besov space can be characterized by the size of the new frame coefficients $\{\langle \tilde{\mathfrak{F}}^{-1}[\tilde{\phi}_\mu], f\rangle\}_{\mu \in \Lambda}$, i.e. if the dual system $\{\tilde{\mathfrak{F}}^{-1}[\tilde{\phi}_\mu]\}_{\mu \in \Lambda}$ is a frame itself. Especially, new frames are discussed, which are built by an approximation of frames stemming from a smooth partition of unity in Fourier space similar to that introduced in Section 2.3. Here, let

$$\{\varphi_{\mathbf{j}}(\mathbf{x}) := \varphi(\mathbf{x} - \mathbf{j})\}_{\mathbf{j} \in \mathbb{Z}^D} \cup \bigcup_{\mathbf{z} \in \mathcal{Z}} \left\{\psi_{l,\mathbf{j}}^{[\mathbf{z}]}(\mathbf{x}) := 2^{\frac{Dl}{2}} \psi^{[\mathbf{z}]}(2^l \mathbf{x} - \mathbf{j})\right\}_{l \in \mathbb{N}_0, \mathbf{j} \in \mathbb{Z}^D}$$

be a frame for $\mathcal{L}^2(\mathbb{R}^D)$ formed by dilation and translation of a generating function $\varphi \in C^m(\mathbb{R}^D)$ and wavelet-like functions $\psi^{[\mathbf{z}]} \in C^m(\mathbb{R}^D)$, $\mathbf{z} \in \mathcal{Z} := \{0,1\}^n \setminus \mathbf{0}$, which also fulfill the following standard assumptions

$$|D^{\mathbf{a}}\varphi(\mathbf{x})| \leq C(1 + |\vec{x}|_2)^{-m_1}, \quad |\mathbf{a}|_1 \leq m_2,$$
$$\left|D^{\mathbf{a}}\psi^{[\mathbf{z}]}(\mathbf{x})\right| \leq C(1 + |\vec{x}|_2)^{-m_1}, \quad |\mathbf{a}|_1 \leq m_2, \mathbf{z} \in \mathcal{Z},$$
$$\int_{\mathbb{R}^D} \mathbf{x}^{\mathbf{a}} \psi^{[\mathbf{z}]} = 0, \quad |\mathbf{a}|_1 < m$$

with appropriate values for the parameters $m_1, m_2, m \in \mathbb{N}$. Note that the behavior of the decay and the number m of vanishing moments is related to the approximation properties of a wavelet; see e.g. [130]. Let further

$$\{\tilde{\varphi}_{\mathbf{j}} := \tilde{\varphi}(\mathbf{x} - \mathbf{j})\}_{\mathbf{j} \in \mathbb{Z}^D} \cup \bigcup_{\mathbf{z} \in \mathcal{Z}} \left\{\tilde{\psi}_{l,\mathbf{j}}^{[\mathbf{z}]} := 2^{\frac{Dl}{2}} \tilde{\psi}^{[\mathbf{z}]}(2^l \mathbf{x} - \mathbf{j})\right\}_{l \in \mathbb{N}_0, \mathbf{j} \in \mathbb{Z}^D} \qquad (4.20)$$

be a system formed by dilation and translation of a generating function $\tilde{\varphi} \in C^m(\mathbb{R}^D)$ and functions $\tilde{\psi}^{[\mathbf{z}]} \in C^m(\mathbb{R}^D)$, $\mathbf{z} \in \mathcal{Z}$, which satisfy the following approximation conditions

$$|D^{\mathbf{a}}\tilde{\varphi}(\mathbf{x}) - D^{\mathbf{a}}\varphi(\mathbf{x})| \leq \epsilon(1 + |\vec{x}|_2)^{-m_1}, \quad |\mathbf{a}|_1 \leq m_2,$$
$$\left|D^{\mathbf{a}}\tilde{\psi}^{[\mathbf{z}]}(\mathbf{x}) - D^{\mathbf{a}}\psi^{[\mathbf{z}]}(\mathbf{x})\right| \leq \epsilon(1 + |\vec{x}|_2)^{-m_1}, \quad |\mathbf{a}|_1 \leq m_2, \mathbf{z} \in \mathcal{Z},$$
$$\int_{\mathbb{R}^D} \mathbf{x}^{\mathbf{a}} \tilde{\psi}^{[\mathbf{z}]} = 0, \quad |\mathbf{a}|_1 < m.$$

In [119] it is shown that for sufficiently small $\epsilon > 0$ and sufficiently large m_1, m_2, m, the system (4.20) also constitutes a frame for $\mathcal{L}^2(\mathbb{R}^D)$. Furthermore, certain Besov and Triebel-Lizorkin spaces can be characterized by the size of the corresponding frame coefficients. In particular, in [118] new basis functions are considered, which are constructed by an approximation in terms of shifted and dilated isotropic Gaussians, i.e. $\varphi(\mathbf{x}) = e^{-|\mathbf{x}|_2^2}$, of an isotropic one-particle Meyer-wavelet basis. Moreover, it is discussed how this newly constructed basis can be utilized in the framework of a nonlinear best M-term approximation.

In the following motivated by the approach based on the principle of small perturbations, we introduce certain multiscale Gaussian frames and shortly discuss their application to the electronic Schrödinger equation in the framework of a generalized hyperbolic cross approach similar to that considered in Section 2.3. Note that in Section 2.3.1 we considered multiscale systems $\mathcal{B}_{V_{L;J}^{T;R}}$ for spanning general sparse grid spaces as in (2.50). These systems are particularly based on a partition of unity on $(\mathbb{R}^D)^N$, which is given by tensor products of a partition of unity on $\mathbb{R}^D$; compare (2.28) and (2.29).

4.2.2 One-particle multiscale Gaussian frame

Here, we consider certain families of functions in the one-particle space $\mathcal{L}^2(\mathbb{R}^D)$. With the help of continuous square integrable one-particle functions φ and $\psi^{[\mathbf{z}]}$ for $\mathbf{z} \in \mathcal{Z} := \{0,1\}^D \setminus \mathbf{0}$ we introduce a family of functions by

$$\left\{\varphi_{\mathbf{j}} : \mathbf{j} \in \mathbb{Z}^D\right\} \cup \bigcup_{\mathbf{z} \in \mathcal{Z}} \left\{\psi^{[\mathbf{z}]}_{l,\mathbf{j}} : l \in \mathbb{N}_0, \mathbf{j} \in \mathbb{Z}^D\right\}, \qquad (4.21)$$

where we set

$$\varphi_{\mathbf{j}}(\mathbf{x}) := c^{\frac{D}{2}} \varphi(c\mathbf{x} - b\mathbf{j}), \quad \psi^{[\mathbf{z}]}_{l,\mathbf{j}} := (c2^l)^{\frac{D}{2}} \psi^{[\mathbf{z}]}(c2^l \mathbf{x} - b\mathbf{j})$$

with a scaling constant $c > 0$ and a dilatation constant $b > 0$. For such a family of one-particle functions we estimate the corresponding frame bounds analogously to the analysis in [37]. Here, to shorten notation we introduce

$$\tilde{\alpha}[f] := \sum_{\mathbf{j} \in \mathbb{Z}^D} |\langle \varphi_{\mathbf{j}}, f \rangle|^2 + \sum_{\mathbf{z} \in \mathcal{Z}, l \in \mathbb{N}_0, \mathbf{j} \in \mathbb{Z}^D} |\langle \psi^{(\mathbf{z})}_{l,\mathbf{j}}, f \rangle|^2,$$

and we set $\gamma_l := c2^l$ for $l \in \mathbb{N}_0$ and $\tilde{b} := \frac{2\pi}{b}$. We obtain

$$\tilde{\alpha}[f] = \sum_{\mathbf{j} \in \mathbb{Z}^D} |\langle \hat{\varphi}_{\mathbf{j}}, \hat{f} \rangle|^2 + \sum_{\mathbf{z} \in \mathcal{Z}, l \in \mathbb{N}_0, \mathbf{j} \in \mathbb{Z}^D} |\langle \hat{\psi}^{(\mathbf{z})}_{l,\mathbf{j}}, \hat{f} \rangle|^2$$

$$= \sum_{\mathbf{j} \in \mathbb{Z}^D} \gamma_0^{-D} \left| \int_{[0,\tilde{b}\gamma_0]^D} e^{ib\gamma_0^{-1}\mathbf{j}^T\mathbf{k}} \sum_{\mathbf{j}' \in \mathbb{Z}^D} \left(\hat{\varphi}(\gamma_0^{-1}\mathbf{k} + \tilde{b}\mathbf{j}')\right)^* \hat{f}(\mathbf{k} + \tilde{b}\mathbf{j}'\gamma_0) \right|^2 d\mathbf{k}$$

$$+ \sum_{\mathbf{z} \in \mathcal{Z}, l \in \mathbb{N}_0, \mathbf{j} \in \mathbb{Z}^D} \gamma_l^{-D} \left| \int_{[0,\tilde{b}\gamma_l]^D} e^{ib\gamma_l^{-1}\mathbf{j}^T\mathbf{k}} \sum_{\mathbf{j}' \in \mathbb{Z}^D} \left(\hat{\psi}^{[\mathbf{z}]}(\gamma_l^{-1}\mathbf{k} + \tilde{b}\mathbf{j}')\right)^* \hat{f}(\mathbf{k} + \tilde{b}\mathbf{j}'\gamma_l) \right|^2 d\mathbf{k}$$

$$= \tilde{b}^D \left(\sum_{\mathbf{j}' \in \mathbb{Z}^D} \int_{\mathbb{R}^D} \hat{f}(\mathbf{k}) \left(\hat{f}(\mathbf{k} + \tilde{b}\mathbf{j}'\gamma_0)\right)^* \left(\hat{\varphi}(\gamma_0^{-1}\mathbf{k})\right)^* \hat{\varphi}(\gamma_0^{-1}\mathbf{k} + \tilde{b}\mathbf{j}') d\mathbf{k} \right. \qquad (4.22)$$

$$+ \left. \sum_{\mathbf{z} \in \mathcal{Z}, l \in \mathbb{N}_0, \mathbf{j}' \in \mathbb{Z}^D} \int_{\mathbb{R}^D} \hat{f}(\mathbf{k}) \left(\hat{f}(\mathbf{k} + \tilde{b}\mathbf{j}'\gamma_l)\right)^* \left(\hat{\psi}^{[\mathbf{z}]}(\gamma_l^{-1}\mathbf{k})\right)^* \hat{\psi}^{[\mathbf{z}]}(\gamma_l^{-1}\mathbf{k} + \tilde{b}\mathbf{j}') d\mathbf{k} \right)$$

$$= \tilde{b}^D \int_{\mathbb{R}^D} |\hat{f}(\mathbf{k})|^2 \alpha(\mathbf{k}) \, d\mathbf{k} + \mathcal{R}[f],$$

where we set

$$\alpha(\mathbf{k}) := |\hat{\varphi}(\gamma_0^{-1}\mathbf{k})|^2 + \sum_{\mathbf{z} \in \mathcal{Z}, l \in \mathbb{N}_0} |\hat{\psi}^{[\mathbf{z}]}(\gamma_l^{-1}\mathbf{k})|^2.$$

and $\mathcal{R}[f]$ denotes all the remaining terms. Here, with the help of the Cauchy-Schwarz inequality, a bound for $|\mathcal{R}[f]|$ can be deduced in the form

$$
\begin{aligned}
|\mathcal{R}[f]| = \tilde{b}^D \bigg| &\sum_{\mathbf{j}' \in \mathbb{Z}^D \setminus \mathbf{0}} \int_{\mathbb{R}^D} \hat{f}(\mathbf{k}) \left(\hat{f}(\mathbf{k} + \tilde{b}\mathbf{j}'\gamma_0)\right)^* \left(\hat{\varphi}(\gamma_0^{-1}\mathbf{k})\right)^* \hat{\varphi}(\gamma_0^{-1}\mathbf{k} + \tilde{b}\mathbf{j}') \, d\mathbf{k} \\
&+ \sum_{\mathbf{z} \in \mathcal{Z}, l \in \mathbb{N}_0} \sum_{\mathbf{j}' \in \mathbb{Z}^D \setminus \mathbf{0}} \int_{\mathbb{R}^D} \hat{f}(\mathbf{k}) \left(\hat{f}(\mathbf{k} + \tilde{b}\mathbf{j}'\gamma_l)\right)^* \left(\hat{\psi}^{[\mathbf{z}]}(\gamma_l^{-1}\mathbf{k})\right)^* \hat{\psi}^{[\mathbf{z}]}(\gamma_l^{-1}\mathbf{k} + \tilde{b}\mathbf{j}') \, d\mathbf{k} \bigg| \\
\leq \tilde{b}^D & \sum_{\mathbf{j}' \in \mathbb{Z}^D \setminus \mathbf{0}} \left(\tilde{\beta}[f](\tilde{b}\mathbf{j}')\tilde{\beta}[f](-\tilde{b}\mathbf{j}')\right)^{\frac{1}{2}} \\
\leq \tilde{b}^D & \|f\|_{\mathcal{L}^2}^2 \sum_{\mathbf{j}' \in \mathbb{Z}^D \setminus \mathbf{0}} \left(\beta(\tilde{b}\mathbf{j}')\beta(-\tilde{b}\mathbf{j}')\right)^{\frac{1}{2}},
\end{aligned}
\tag{4.23}
$$

where to shorten notation we set

$$
\tilde{\beta}[f](\mathbf{k}') := \int_{\mathbb{R}^D} |\hat{f}(\mathbf{k})|^2 \Big(|\hat{\varphi}(\gamma_0^{-1}\mathbf{k})||\hat{\varphi}(\gamma_0^{-1}\mathbf{k} + \mathbf{k}')| \\
+ \sum_{\mathbf{z} \in \mathcal{Z}, l \in \mathbb{N}_0} |\hat{\psi}^{[\mathbf{z}]}(\gamma_l^{-1}\mathbf{k})||\hat{\psi}^{[\mathbf{z}]}(\gamma_l^{-1}\mathbf{k} + \mathbf{k}')| \Big) \, d\mathbf{k}
$$

and

$$
\beta(\mathbf{k}') := \sup_{\mathbf{k} \in \mathbb{R}^D} \Big(|\hat{\varphi}(\gamma_0^{-1}\mathbf{k})||\hat{\varphi}(\gamma_0^{-1}\mathbf{k} + \mathbf{k}')| + \sum_{\mathbf{z} \in \mathcal{Z}, l \in \mathbb{N}_0} |\hat{\psi}^{[\mathbf{z}]}(\gamma_l^{-1}\mathbf{k})||\hat{\psi}^{[\mathbf{z}]}(\gamma_l^{-1}\mathbf{k} + \mathbf{k}')|\Big).
$$

Then, with the help of (4.22) and (4.23) we obtain the inequalities

$$
\inf_{f \in \mathcal{L}^2(\mathbb{R}^D) \setminus 0} \|f\|_{\mathcal{L}^2}^{-2} \tilde{\alpha}[f] \geq \left(\frac{2\pi}{b}\right)^D \left(\inf_{\mathbf{k} \in \mathbb{R}^D} \alpha(\mathbf{k}) - \sum_{\mathbf{j}' \in \mathbb{Z}^D \setminus \mathbf{0}} \left(\beta(\tfrac{2\pi}{b}\mathbf{j}')\beta(-\tfrac{2\pi}{b}\mathbf{j}')\right)^{\frac{1}{2}}\right), \tag{4.24}
$$

$$
\sup_{f \in \mathcal{L}^2(\mathbb{R}^D) \setminus 0} \|f\|_{\mathcal{L}^2}^{-2} \tilde{\alpha}[f] \leq \left(\frac{2\pi}{b}\right)^D \left(\sup_{\mathbf{k} \in \mathbb{R}^D} \alpha(\mathbf{k}) + \sum_{\mathbf{j}' \in \mathbb{Z}^D \setminus \mathbf{0}} \left(\beta(\tfrac{2\pi}{b}\mathbf{j}')\beta(-\tfrac{2\pi}{b}\mathbf{j}')\right)^{\frac{1}{2}}\right) \tag{4.25}
$$

with which the following lemma can be proved in a similar way to [37]:

Lemma 4.1. *Let φ and $\psi^{[\mathbf{z}]}$ for $\mathbf{z} \in \mathcal{Z}$ be continuous one-particle functions such that*

$$
\inf_{\mathbf{k} \in \mathbb{R}^D} \alpha(\mathbf{k}) > 0, \quad \sup_{\mathbf{k} \in \mathbb{R}^D} \alpha(\mathbf{k}) < \infty
$$

and such that $\beta(\mathbf{k}')$ decays at least as fast as $(1 + |\mathbf{k}'|_2)^{-(1+\epsilon)}$ with $\epsilon > 0$. Then there exists $b' > 0$ such that for all $\beta < \beta'$ the family of functions given in (4.21) forms a frame and a possible pair of frame bounds C_1, C_2 in (4.18) is given by the right hand sides of (4.24) and (4.25), respectively.

Proof. Due to the decay of β there exists a b' so that $\sum_{\mathbf{j}' \in \mathbb{Z}^D \setminus \mathbf{0}} \left(\beta\left(\tfrac{2\pi}{b}\mathbf{j}'\right)\beta\left(-\tfrac{2\pi}{b}\mathbf{j}'\right)\right)^{\frac{1}{2}} <$ $\inf_{\mathbf{k} \in \mathbb{R}^D} \alpha(\mathbf{k})$ for $b < b'$. Then, with the boundedness of α, it follows that the right hand sides of (4.24) and (4.25) are strictly positive and bounded, which leads to the desired result. $\square$

Let us remark that Lemma 4.1 just gives a sufficient but not a necessary condition for system (4.21) to constitute a frame. For example, for a one-dimensional Meyer basis with $c = b = 1$, as in Appendix B.2, the frame constants are equal to 1 since it is an orthonormal basis. However, the right hand side of (4.24) is not strictly positive in that case. Nevertheless, for $c = 1$, $b = 1/4$ we approximately computed estimation values $C_1 \approx 4.0$, $C_2 \approx 4.0$, $C_1/C_1 \approx 1.0$ for the frame bounds and hence system (4.21) then is a frame due to Lemma 4.1.

Gaussian frame of second order

We introduce a generating function in $\mathcal{L}^2(\mathbb{R}^D)$ by

$$\varphi_\sigma : \mathbb{R}^D \to \mathbb{R} : \mathbf{x} \mapsto \left(\bigotimes_{d=1}^{D} g_\sigma\right)(\mathbf{x}), \qquad (4.26)$$

where g_σ denotes an $\mathcal{L}^2$-normalized one-dimensional Gaussian (4.16). By dilation and translation we then define the functions

$$\varphi_{\sigma,c,l,\mathbf{j}}(\mathbf{x}) := \left(c2^l\right)^{\frac{D}{2}} \varphi_\sigma(c2^l\mathbf{x} - \mathbf{j}) \qquad (4.27)$$

for $c > 0$, $l \in \mathbb{N}_0$ and $\mathbf{j} \in \mathbb{Z}^D$. In particular, $\varphi_{\sigma,c,l,\mathbf{j}}$ is $\mathcal{L}^2$-normalized and its zeroth moment does not vanish, i.e. $\int_{\mathbb{R}^D} \varphi_{\sigma,c,l,\mathbf{j}}(\mathbf{x})\,d\mathbf{x} = \left(\frac{2\sigma\sqrt{\pi}}{c2^l}\right)^{\frac{1}{2}}$. Furthermore, the function $\varphi_{\sigma,c,l,\mathbf{j}}$ fulfills the second order moment condition (4.19). Moreover, we introduce a wavelet-like function in terms of generating functions of two scales by

$$\psi_\sigma(\mathbf{x}) := C_{\psi_\sigma}\left(\varphi_{\frac{1}{2}\sigma}(\mathbf{x}) - \gamma\varphi_\sigma(\mathbf{x})\right),$$

with the normalization constant $C_{\psi_\sigma} = (1 - \frac{16}{25}\gamma\sqrt{5} + \gamma^2)^{-\frac{1}{2}}$; compare also [128]. In particular, we set $\gamma = 2^{-\frac{D}{2}}$ and define the functions

$$\psi^{[\mathbf{z}]}_{\sigma,c,l,\mathbf{j}}(\mathbf{x}) := \left(c2^l\right)^{\frac{D}{2}} \psi_\sigma(c2^l\mathbf{x} - \mathbf{j} - \tfrac{1}{2}\mathbf{z}) \qquad (4.28)$$

by dilation and translation for $c > 0$, $l \in \mathbb{N}_0$, $\mathbf{j} \in \mathbb{Z}^D$ and $\mathbf{z} \in \mathcal{Z}$. Here, $\psi^{[\mathbf{z}]}_{\sigma,c,l,\mathbf{j}}$ is $\mathcal{L}^2$-normalized and its zeroth and first moments vanish, i.e. $\int_{\mathbb{R}^D} \mathbf{x}^\mathbf{a}\psi^{[\mathbf{z}]}_{\sigma,c,l,\mathbf{j}}(\mathbf{x})\,d\mathbf{x} = 0$ for all $\mathbf{a} \in \mathbb{N}_0^D$, $0 \leq |\mathbf{a}|_1 < 2$.

Examples for the functions φ_σ and ψ_σ are depicted in Figure 4.1 (top). For the system of one-particle functions

$$\mathfrak{b}_{\sigma,c} := \left\{\varphi_{\sigma,c,0,\mathbf{j}} : \mathbf{j} \in \mathbb{Z}^D\right\} \cup \bigcup_{\mathbf{z}\in\mathcal{Z}} \left\{\psi^{[\mathbf{z}]}_{\sigma,c,l,\mathbf{j}} : l \in \mathbb{N}_0, \mathbf{j} \in \mathbb{Z}^D\right\} \qquad (4.29)$$

we compute estimates for the corresponding frame bounds in Lemma 4.1. In the case of $c = b = 1$ we obtain the estimated values $C_1 \approx 1183.88$, $C_2 \approx 2850.98$, $C_1/C_2 \approx 2.41$ for $\sigma = 4$ and hence, $\mathfrak{b}_{4,1}$ constitutes a frame. Note that we have $\varphi_{\sigma,c,l,\mathbf{j}}(\mathbf{x}) = \left(\frac{c}{\sigma}2^l\right)^{\frac{D}{2}} \varphi_1(\frac{c}{\sigma}2^l\mathbf{x} - \frac{1}{\sigma}\mathbf{j})$ and $\psi^{[\mathbf{z}]}_{\sigma,c,l,\mathbf{j}}(\mathbf{x}) = \left(\frac{c}{\sigma}2^l\right)^{\frac{D}{2}} \psi_1(\frac{c}{\sigma}2^l\mathbf{x} - \frac{1}{\sigma}\mathbf{j} - \frac{1}{2\sigma}\mathbf{z})$. Therefore, the reciprocal of the deviation parameter here takes the role of parameter b in Lemma 4.1, i.e. $b = \frac{1}{\sigma}$.

4 Numerical Methods

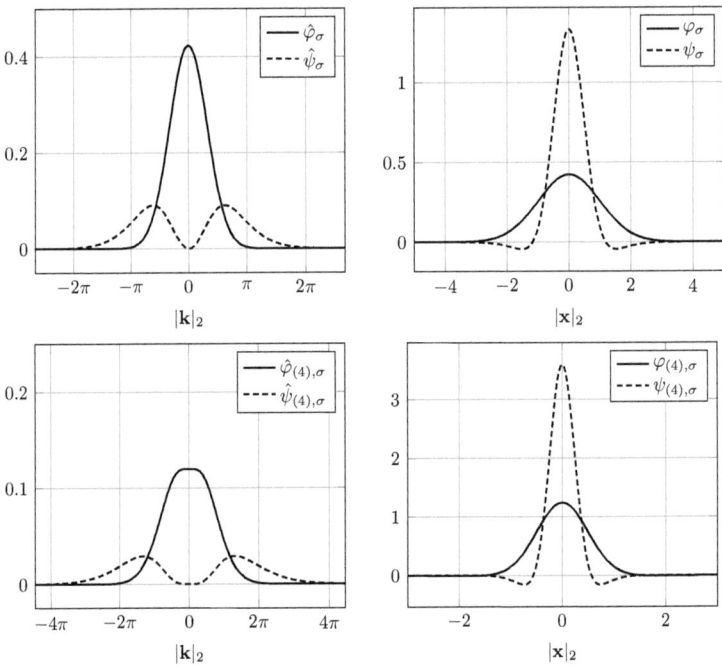

Figure 4.1: Top: φ_σ and ψ_σ from (4.27) and (4.28), respectively, in Fourier space (left) and real space (right) for $\sigma = 1$, $D = 3$. Bottom: $\varphi_{(4),\sigma}$ and $\psi_{(4),\sigma}$ from (4.30) and (4.31), respectively, in Fourier space (left) and real space (right) for $\sigma = 1$, $D = 3$.

Gaussian frame of fourth order

Generating functions which satisfy moment conditions (4.19) of higher order than two can be constructed by several general schemes; see e.g. [135]. For example, with the help of the Vandermonde determinant and Cramer's rule higher order generating functions in terms of linear combinations of Gaussians with different deviation parameters can be deduced. Analogously, also wavelet-like functions which exhibit a higher number of vanishing moments can be constructed. In the following we simply give an example in the case of fourth order.

To this end, we introduce a generating one-particle function in $\mathcal{L}^2(\mathbb{R}^3)$ in the form

$$\varphi_{(4),\sigma} : \mathbb{R}^3 \to \mathbb{R} : \mathbf{x} \mapsto C_{\varphi_{(4),\sigma}} \left(2^{\frac{3}{2}} \varphi_\sigma(2\mathbf{x}) - 2^{-\frac{7}{2}} \varphi_\sigma(\mathbf{x}) \right) \quad (4.30)$$

with a normalization constant of value $C_{\varphi_{(4),\sigma}} = \frac{80}{\sqrt{6450 - 256\sqrt{2}\sqrt{5}}}$. By dilation and translation we then define functions

$$\varphi_{(4),\sigma,c,l,\mathbf{j}}(\mathbf{x}) := \left(c2^l \right)^{\frac{3}{2}} \varphi_\sigma(c2^l \mathbf{x} - \mathbf{j})$$

for $c > 0$, $l \in \mathbb{N}_0$ and $\mathbf{j} \in \mathbb{Z}^3$. Here, $\varphi_{\sigma,c,l,\mathbf{j}}$ is $\mathcal{L}^2$-normalized and fulfills the fourth order moment condition (4.19). Moreover, we define a wavelet-like function

$$\psi_{(4),\sigma}(\mathbf{x}) := C_{\psi_{(4),\sigma}} \left(\varphi_{(4),\frac{1}{2}\sigma}(\mathbf{x}) - 2^{-\frac{3}{2}} \varphi_{(4),\sigma}(\mathbf{x}) \right) \qquad (4.31)$$

with the normalization constant $C_{\psi_{(4),\sigma}} = \dfrac{544\sqrt{5}}{\sqrt{1770125 - 305184\sqrt{2}\sqrt{5} + 5120\sqrt{2}\sqrt{17}}}$. Again, by dilation and translation we define functions

$$\psi^{[\mathbf{z}]}_{(4),\sigma,c,l,\mathbf{j}}(\mathbf{x}) := \left(c2^l\right)^{\frac{3}{2}} \psi_\sigma(c2^l \mathbf{x} - \mathbf{j} - \tfrac{1}{2}\mathbf{z})$$

for $c > 0$, $l \in \mathbb{N}_0$, $\mathbf{j} \in \mathbb{Z}^3$ and $\mathbf{z} \in \{0,1\}^3 \setminus \mathbf{0}$. Here, $\psi^{[\mathbf{z}]}_{(4),\sigma,c,l,\mathbf{j}}$ is $\mathcal{L}^2$-normalized and its zeroth, first, second and third moments vanish, i.e. $\int_{\mathbb{R}^3} \mathbf{x}^{\mathbf{a}} \psi^{[\mathbf{z}]}_{(4),\sigma,c,l,\mathbf{j}}(\mathbf{x})\, d\mathbf{x} = 0$ for all $\mathbf{a} \in \mathbb{N}_0^3$, $0 \leq |\mathbf{a}|_1 < 4$. Note that $\psi_{(4),\sigma}$ can be written in terms of the second order generating function (4.26) in the form

$$\psi_{(4),\sigma}(\mathbf{x}) = C_{\psi_{(4),\sigma}} \left(\varphi_{\frac{1}{4}\sigma}(\mathbf{x}) - (2^{-\frac{7}{2}} + 2^{-\frac{3}{2}}) \varphi_{\frac{1}{2}\sigma}(\mathbf{x}) + 2^{-5} \varphi_\sigma(\mathbf{x}) \right).$$

Examples of the functions $\varphi_{(4),\sigma}$ and $\psi_{(4),\sigma}$ are depicted in Figure 4.1 (bottom). For the sytem of one-particle functions

$$\mathfrak{b}_{(4),\sigma,c} := \left\{ \varphi_{(4),\sigma,c,0,\mathbf{j}} : \mathbf{j} \in \mathbb{Z}^D \right\} \cup \bigcup_{\mathbf{z} \in Z} \left\{ \psi^{[\mathbf{z}]}_{(4),\sigma,c,l,\mathbf{j}} : l \in \mathbb{N}_0, \mathbf{j} \in \mathbb{Z}^D \right\} \qquad (4.32)$$

we compute estimates for corresponding frame bounds from Lemma 4.1. In the case of $c = b = 1$ we obtain the estimated values $C_1 \approx 104.39$, $C_2 \approx 227.62$, $C_1/C_2 \approx 2.18$ for $\sigma = 4$ and hence $\mathfrak{b}_{(4),4,1}$ constitutes a frame. Analogously to the case of system (4.29), the reciprocal of the deviation parameter here takes the role of parameter b in Lemma 4.1, i.e. $b = \frac{1}{\sigma}$.

4.2.3 Many-particle Gaussian frame

In order to obtain a multiscale Gaussian frame for the partially antisymmetric N-particle space, i.e. $\mathcal{L}^2_{(N,M_S)} = \mathfrak{A}^{(N,M_S)}(\mathcal{L}^2((\mathbb{R}^3)^N))$ as in (3.13), we use a tensor product construction analogous to that considered in (4.5) and (4.4). Here, we apply a one-particle multiscale Gaussian frame for $\mathcal{L}^2(\mathbb{R}^D)$ from Section 4.2.2, e.g. $\mathfrak{b}_{\sigma,c}$ or $\mathfrak{b}_{(4),\sigma,c}$ given by (4.29) and (4.32), respectively.[9] In the following the frame for $\mathcal{L}^2_{(N,M_S)}$ which is based on $\mathfrak{b}_{\sigma,c}$ is denoted by $\mathcal{B}^{(N,M_S)}_{\sigma,c}$ and the frame for $\mathcal{L}^2_{(N,M_S)}$ which is based on $\mathfrak{b}_{(4),\sigma,c}$ is denoted by $\mathcal{B}^{(N,M_S)}_{(4),\sigma,c}$.

Note that in the framework of the Galerkin approximation with respect to a finite subset of $\mathcal{B}^{(N,M_S)}_{\sigma,c}$ or $\mathcal{B}^{(N,M_S)}_{(4),\sigma,c}$ all entries in the system matrices (4.2) can be computed by the Löwdin rules considered in Section 4.1.1 and with the help of the integral formulae given in Appendix C.1.

---

[9] With respect to partially antisymmetric spaces see also Section 2.5 and particularly compare (2.28) and (2.29).

4.3 Approximation with finite-order weights

In the following we discuss the application of a higher-order particle-wise decomposition like in Section 2.4.1 in the framework of the computation of an approximate solution of the weak electronic Schrödinger equation (3.16). Here, for fixed (N, M_S) we consider a subspace splitting analogous to (2.90) of the partially antisymmetric N-particle space $\mathcal{L}^2_{(N,M_S)}$; compare Section 2.5.2. In particular, in accordance with (2.93) we perform a particle-wise decomposition with respect to normalized one-particle functions related to a rank-1 approximation.

Furthermore, we introduce new finite-dimensional weighted approximation spaces. In particular, these types of approximation spaces include spaces of finite-order weights (cf. Section 2.4.2) as well as so-called configuration interaction (CI) approximation spaces, which are well-known in quantum chemistry [170].

4.3.1 Particle-wise decomposition

Let $\psi_1, \ldots, \psi_N$ be $\mathcal{L}^2$-normalized one-particle functions, where $\psi_1, \ldots, \psi_{N_\uparrow}$ as well as $\psi_{N_\uparrow+1}, \ldots, \psi_N$ are linear independent. For a shorter notation we set $\mathcal{N}_\uparrow := \{1, \ldots, N_\uparrow\} \subset \mathcal{N}$ and $\mathcal{N}_\downarrow := \{N_\uparrow+1, \ldots, N\} \subset \mathcal{N}$. Let further the subspace $U_\uparrow \subset \mathcal{L}^2(\mathbb{R}^3)$ be spanned by $\{\psi_p\}_{p \in \mathcal{N}_\uparrow}$, and let the subspace $W_\uparrow \subset \mathcal{L}^2(\mathbb{R}^3)$ have a direct sum decomposition $\mathcal{L}^2(\mathbb{R}^3) = U_\uparrow \oplus W_\uparrow$. Analogously, let the subspace $U_\downarrow \subset \mathcal{L}^2(\mathbb{R}^3)$ be spanned by $\{\psi_p\}_{p \in \mathcal{N}_\downarrow}$ and let the subspace $W_\downarrow \subset \mathcal{L}^2(\mathbb{R}^3)$ have a direct sum decomposition $\mathcal{L}^2(\mathbb{R}^3) = U_\downarrow \oplus W_\downarrow$. Then, the subspace splitting

$$\begin{aligned}
\mathcal{L}^2_{(N,M_S)} &= \mathfrak{A}_{N_\uparrow}\left(\bigotimes_{p \in \mathcal{N}_\uparrow}(U_\uparrow \oplus W_\uparrow)\right) \otimes \mathfrak{A}_{N_\downarrow}\left(\bigotimes_{p \in \mathcal{N}_\downarrow}(U_\downarrow \oplus W_\downarrow)\right) \\
&= \bigoplus_{\mu=0}^{N} \underbrace{\bigoplus_{u \subset \mathcal{N}, |u|=\mu} \mathcal{W}^{(N,M_S)}_u}_{=: \mathcal{V}^{\mathfrak{A}(N,M_S)}_\mu}
\end{aligned} \qquad (4.33)$$

follows, where we set similar to (2.91)

$$\mathcal{W}^{(N,M_S)}_u := \mathfrak{A}^{(N,M_S)}\left(\bigotimes_{p=1}^{N} \mathcal{W}^{(N,M_S)}_{u,(p)}\right), \quad \mathcal{W}^{(N,M_S)}_{u,(p)} := \begin{cases} \operatorname{span}\{\psi_p\} & \text{for } p \in \mathcal{N} \setminus u, \\ W_\uparrow & \text{for } p \in u \cap \mathcal{N}_\uparrow, \\ W_\downarrow & \text{for } p \in u \cap \mathcal{N}_\downarrow. \end{cases} \qquad (4.34)$$

Correspondingly, similar to (2.92), any function $f \in \mathcal{L}^2_{(N,M_S)}$ can be decomposed as

$$f = \sum_{u \subset \mathcal{N}} \mathfrak{A}^{(N,M_S)}(F_u) = \sum_{u \subset \mathcal{N}} \mathfrak{A}^{(N,M_S)}\left(\vec{y} \mapsto f_u(\vec{y}_u) \prod_{p \in \mathcal{N} \setminus u} \psi_p(\mathbf{y}_p)\right)$$

with the help of appropriate linear projections such that $\mathfrak{A}^{(N,M_S)} F_u \in \mathcal{W}^{(N,M_S)}_u$ and $f_u \in \bigotimes_{p \in u} \mathcal{W}^{(N,M_S)}_{u,(p)}$. Moreover, in the case of orthogonal direct sums $U_\uparrow \oplus W_\uparrow$, $U_\downarrow \oplus W_\downarrow$ and orthonormal $\{\psi_p\}_{p \in \mathcal{N}_\uparrow}$, $\{\psi_p\}_{p \in \mathcal{N}_\downarrow}$, the orthogonality relation $\langle F_u, F_v \rangle = 0$ holds for all $u \neq v$. We already noted in Section 2.5.2 that a similar type of a particle-wise decomposition was introduced by Sinanoğlu in quantum chemistry for the analysis of many-electron wave functions [164]. Here, Hartree-Fock orbitals are suggested as one-particle functions $\{\psi_p\}_{p \in \mathcal{N}}$.

The Hartree-Fock approximation is given analogously to (2.93) as the best rank-1 approximation of the lowest state of the electronic Schrödinger equation (3.16), i.e. the so-called Hartree-Fock orbitals $\psi_1^{HF}, \ldots, \psi_N^{HF} \in \mathcal{H}^1(\mathbb{R}^3)$ minimize the Hartree-Fock energy functional[10]

$$\frac{\left\langle \tilde{\mathfrak{A}}^{(N,M_s)}[\vec{\psi}^{HF}], H\tilde{\mathfrak{A}}^{(N,M_s)}[\vec{\psi}^{HF}]\right\rangle}{\left\langle \tilde{\mathfrak{A}}^{(N,M_s)}[\vec{\psi}^{HF}], \tilde{\mathfrak{A}}^{(N,M_s)}[\vec{\psi}^{HF}]\right\rangle} = \inf_{\Psi_{(N,M_S)}^{HF} \in \mathcal{V}_{(N,M_S)}^{HF}} \left\langle \Psi_{(N,M_S)}^{HF}, H\Psi_{(N,M_S)}^{HF}\right\rangle, \quad (4.35)$$

where

$$\mathcal{V}_{(N,M_S)}^{HF} := \left\{ \tilde{\mathfrak{A}}^{(N,M_s)}[\vec{\psi}] \ : \ \psi_1, \ldots, \psi_N \in \mathcal{H}^1(\mathbb{R}^3), \|\tilde{\mathfrak{A}}^{(N,M_S)}[\vec{\psi}]\| = 1 \right\}.$$

Here, the minimization of the energy functional $\left\langle \Psi_{(N,M_S)}^{HF}, H\Psi_{(N,M_S)}^{HF}\right\rangle$ in $\mathcal{V}_{(N,M_S)}^{HF}$ leads to an Euler-Lagrange equation, which then corresponds to the so-called general unrestricted Hartree-Fock eigenvalue equation [121]. Its discretization leads to the well-known Roothaan-Hall equations in the closed-shell case, i.e. $N_\uparrow = N_\downarrow$, and to the well-known Pople-Nesbet equations in the open-shell case, i.e. $N_\uparrow \neq N_\downarrow$; compare [170].

The Roothaan-Hall equations can be written in a form resembling a generalized nonlinear eigenvalue problem, whereas the Pople-Nesbet equations can be written as a coupled pair of certain Roothaan-Hall equations. In particular, since the Hartree-Fock energy functional is invariant with respect to unitary transformations of spin-up and spin-down orbitals, the spin-up Hartree-Fock orbitals $\{\psi_p^{HF}\}_{p \in \mathcal{N}_\uparrow}$ as well as the spin-down Hartree-Fock orbitals $\{\psi_p^{HF}\}_{p \in \mathcal{N}_\downarrow}$ can be assumed to be orthonormal. For more details of the Hartree-Fock approximation see [121, 170], for example.

In this thesis we apply non-orthogonal approximate Hartree-Fock orbitals. For its computation we perform a minimization of the energy functional

$$\inf_{\tilde{\Psi}_{(N,M_S)}^{HF} \in \tilde{\mathcal{V}}_{(N,M_S)}^{HF}} \left\langle \tilde{\Psi}_{(N,M_S)}^{HF}, H\tilde{\Psi}_{(N,M_S)}^{HF}\right\rangle \quad (4.36)$$

by a multidimensional minimization algorithm without derivatives for reasons of simplicity, i.e. we employ the simplex algorithm of Nelder and Mead as implemented in the GNU Scientific Library (GSL) [59, 144]. Here, $\tilde{\mathcal{V}}_{(N,M_S)}^{HF}$ is a subset of $\mathcal{V}_{(N,M_S)}^{HF}$, where only determinants $\tilde{\mathfrak{A}}^{(N,M_s)}[\vec{\psi}]$, which are built from certain types of one-particle functions $\tilde{\psi}_1, \ldots, \tilde{\psi}_N \in C^\infty(\mathbb{R}^3)$, are taken into account. The respective type of a one-particle function is chosen with regard to accurate representation of a certain occupied atomic or molecular orbital by variation of just a few variables. For the s-type atomic orbitals we use a finite expansion of length Q in terms of isotropic Gaussians (4.17) of different deviation centered at a certain atomic position $\mathbf{R}$, i.e.

$$o_{\{c_1,\ldots,c_Q,\sigma_1,\ldots,\sigma_Q\}}^{(s),\mathbf{R},Q}(\mathbf{x}) := \sum_{q=1}^{Q} c_q G_{\sigma_q}(\mathbf{x} - \mathbf{R}). \quad (4.37)$$

---

[10]Here, like in Section 4.1.1 we use the following notation: $\tilde{\mathfrak{A}}^{(N,M_s)}[\vec{\psi}]$ denotes the partially antisymmetric N-particle function $\tilde{\mathfrak{A}}^{(N,M_S)} \bigotimes_{p=1}^{N} \psi_p$ for a vector of one-particle functions $\vec{\psi} = (\psi_1, \ldots, \psi_N)^T$.

4 Numerical Methods

For p-type atomic orbitals we employ a finite expansion in terms of certain modulated isotropic Gaussians of different deviation, i.e.

$$o^{(p_d),\mathbf{R},Q}_{\{c_1,\ldots,c_Q,c'_1,\ldots,c'_Q,\sigma_1,\ldots,\sigma_Q\}}(\mathbf{x}) := \sum_{q=1}^{Q} c_q \sin(c'_q x_d) G_{\sigma_q}(\mathbf{x}-\mathbf{R}), \qquad (4.38)$$

see, e.g. [3].

To approximately represent certain molecular orbitals we apply linear combinations of s- and p-type orbitals centered at different atomic positions. Note that the set of variation parameters associated with an orbital type is denoted by the set in the lower index in (4.37) and (4.38), respectively. Note further that all necessary one- and two-operator integrals can be given in terms of analytic formulae; see Appendix C. We give the respective specific choice of the type of orbitals for each numerical experiment in Chapter 5.

Now, let us assume that appropriate one-particle functions with $U_\uparrow = \text{span}\{\psi_p : p \in \mathcal{N}_\uparrow\}$ and $U_\downarrow = \text{span}\{\psi_p : p \in \mathcal{N}_\downarrow\}$ are chosen with regard to a fixed electronic Hamiltonian and fixed (N, M_S). For a set of functions in the form $\{\Psi_u \in \mathcal{L}^2((\mathbb{R}^3)^{|u|})\}_{u \subset \mathcal{N}}$, we denote the partially antisymmetric N-particle function by

$$\mathfrak{A}^{(N,M_S)}[\{\Psi_u\}_{u \subset \mathcal{N}}] := \sum_{u \subset \mathcal{N}} \mathfrak{A}^{(N,M_S)} \left(\vec{y} \mapsto \Psi_u(\vec{y}_u) \prod_{p \in \mathcal{N} \setminus u} \psi_p(\mathbf{y}_p) \right).$$

To obtain a particle-wise decomposition of a lowest state associated with the weak electronic Schrödinger equation (3.16), we minimize the electronic energy functional

$$\min_{\substack{\left\{\Psi_u \in \bigotimes_{p \in u} \mathcal{W}^{(N,M_S)}_{u,(p)}\right\}_{u \subset \mathcal{N}} \\ \|\mathfrak{A}^{(N,M_S)}[\{\Psi_u\}_{u \subset \mathcal{N}}]\| = 1}} \left\langle \mathfrak{A}^{(N,M_S)}[\{\Psi_u\}_{u \subset \mathcal{N}}], H\mathfrak{A}^{(N,M_S)}[\{\Psi_u\}_{u \subset \mathcal{N}}] \right\rangle. \qquad (4.39)$$

Here, a minimizing normalized particle-wise decomposition $\mathfrak{A}^{(N,M_S)}[\{\Psi_u\}_{u \subset \mathcal{N}}]$ associated with (4.39) corresponds to a lowest state and the minimum is equal to the minimal energy as in (3.18).

4.3.2 Approximation spaces with weights of finite order

So far, for reasons of simplicity, we used restrictions of the subspace splitting (2.90) in the form (2.95) only. Such a restriction corresponds to a set of finite-order weights of order $q \leq N$. We now consider more general subspaces, which result from a restriction of the splitting in (4.33). To this end, let again $U_\uparrow$ and $U_\downarrow$ be spanned by systems $\{\psi_p\}_{p \in \mathcal{N}_\uparrow}$ and $\{\psi_p\}_{p \in \mathcal{N}_\downarrow}$, and let $W_\uparrow$ and $W_\downarrow$ have the sums $U_\uparrow + W_\uparrow = \mathcal{L}^2(\mathbb{R}^3)$ and $U_\downarrow + W_\downarrow = \mathcal{L}^2(\mathbb{R}^3)$, respectively. In the following a restriction according to the given finite-order weights $\{\gamma_u \geq 0\}_{u \subset \mathcal{N}}$ of a sum of subspaces is denoted by[11]

$$\mathcal{V}^{(N,M_S)}_{\{\gamma_u\}_u} := \sum_{u \subset \mathcal{N}, \gamma_u > 0} \mathcal{W}^{(N,M_S)}_u. \qquad (4.40)$$

---

[11] Here, the finite-order weights $\{\gamma_u\}_{u \subset \mathcal{N}}$ are just used to switch certain subspaces $\mathcal{W}^{(N,M_S)}_u$ on or off. Compare also Section 2.4.2.

4.3 Approximation with finite-order weights

Analogously to (4.39), a particle-wise decomposition corresponding to a sum of subspaces (4.40) can be obtained by

$$\min_{\substack{\{\Psi_u \in \bigotimes_{p \in u} \mathcal{W}_{u,(p)}^{(N,M_S)}\}_{u \subset \mathcal{N}, \gamma_u > 0} \\ \|\mathfrak{A}^{(N,M_S)}[\{\Psi_u\}_{u \subset \mathcal{N}, \gamma_u > 0}]\| = 1}} \left\langle \mathfrak{A}^{(N,M_S)}[\{\Psi_u\}_{u \subset \mathcal{N}, \gamma_u > 0}], H\mathfrak{A}^{(N,M_S)}[\{\Psi_u\}_{u \subset \mathcal{N}, \gamma_u > 0}] \right\rangle. \quad (4.41)$$

Note here that in general a minimizing set $\{\Psi_u\}_{u \subset \mathcal{N}, \gamma_u > 0}$ is not unique.

In the following we introduce finite-dimensional subspaces with regard to an appropriate Galerkin approximation of (4.41). Here, the idea is to construct a finite-dimensional subspace of $\mathcal{V}_{\{\gamma_u\}_u}^{(N,M_S)}$ by choosing a specific finite-dimensional subspace of $\mathcal{W}_u^{(N,M_S)}$ for each u with $\gamma_u > 0$ separately. To this end, let $\{\phi_\nu^\uparrow\}_{\nu \in \mathbb{Z}}$ and $\{\phi_\nu^\downarrow\}_{\nu \in \mathbb{Z}}$ be frames for $W_\uparrow$ and $W_\downarrow$, respectively. Then, for each $u = \{p_1 < \cdots < p_{|u|}\} \subset \mathcal{N}$ with $\gamma_u > 0$, where $\{p_1 < \cdots < p_{n_\uparrow}\} = u \cap \mathcal{N}_\uparrow$ and $\{p_{n_\uparrow+1} < \cdots < p_{|u|}\} = u \cap \mathcal{N}_\downarrow$, we introduce a finite-dimensional subspace $\tilde{\mathcal{W}}_{u,\mathcal{B}_u}^{(N,M_S)}$ of $\mathcal{W}_u^{(N,M_S)}$ by the span of the finite set $\mathcal{B}_u$ of partially antisymmetric N-particle functions. Here, $\mathcal{B}_u$ is chosen as a finite subset of

$$\mathcal{B}_u^{(N,M_S)} := \left\{ \mathfrak{A}^{(N,M_S)}\left(\vec{\mathbf{y}} \mapsto \prod_{q=1}^{n_\uparrow} \phi_{\nu_q}^\uparrow(\mathbf{y}_{p_q}) \prod_{q=n_\uparrow+1}^{|u|} \phi_{\nu_q}^\downarrow(\mathbf{y}_{p_q}) \prod_{p \in \mathcal{N} \setminus u} \psi_p(\mathbf{y}_p) \right) : \right.$$

$$\left. (\nu_1, \ldots, \nu_{|u|})^T \in \mathbb{Z}_u^{(N,M_S)} \right\}, \quad (4.42)$$

where

$$\mathbb{Z}_u^{(N,M_S)} := \left\{ (\nu_1, \ldots, \nu_{|u|})^T \in \mathbb{Z}^{|u|} : \nu_1 < \ldots < \nu_{n_\uparrow}, \nu_{n_\uparrow+1} < \cdots < \nu_{|u|} \right\}.$$

In this way, a finite-dimensional subspace of $\mathcal{V}_{\{\gamma_u\}_u}^{(N,M_S)}$ can be defined by

$$\tilde{\mathcal{V}}_{\{\gamma_u\}_u, \{\mathcal{B}_u\}_u}^{(N,M_S)} := \sum_{u \subset \mathcal{N}, \gamma_u > 0} \tilde{\mathcal{W}}_{u,\mathcal{B}_u}^{(N,M_S)} = \sum_{u \subset \mathcal{N}, \gamma_u > 0} \mathrm{span}(\mathcal{B}_u) \quad (4.43)$$

with the help of a family of systems $\left\{ \mathcal{B}_u \subset \mathcal{B}_u^{(N,M_S)} \right\}_{u \subset \mathcal{N}, \gamma_u > 0}$. Note that we always assume that each system $\mathcal{B}_u \subset \mathcal{B}_u^{(N,M_S)}$ is chosen such that it is linear independent and that we especially have the direct sum decomposition $\tilde{\mathcal{V}}_{\{\gamma_u\}_u, \{\mathcal{B}_u\}_u}^{(N,M_S)} = \bigoplus_{u \subset \mathcal{N}, \gamma_u > 0} \tilde{\mathcal{W}}_{u,\mathcal{B}_u}^{(N,M_S)}$. In particular, in the case of the direct sum decompositions $U_\uparrow \oplus W_\uparrow = \mathcal{L}^2(\mathbb{R}^3)$ and $U_\downarrow \oplus W_\downarrow = \mathcal{L}^2(\mathbb{R}^3)$, we also obtain direct sum decompositions for (4.40) and (4.43), i.e. $\mathcal{V}_{\{\gamma_u\}_u}^{(N,M_S)} = \bigoplus_{u \subset \mathcal{N}, \gamma_u > 0} \mathcal{W}_u^{(N,M_S)}$ and $\tilde{\mathcal{V}}_{\{\gamma_u\}_u, \{\mathcal{B}_u\}_u}^{(N,M_S)} = \bigoplus_{u \subset \mathcal{N}, \gamma_u > 0} \tilde{\mathcal{W}}_{u,\mathcal{B}_u}^{(N,M_S)} = \bigoplus_{u \subset \mathcal{N}, \gamma_u > 0} \mathrm{span}(\mathcal{B}_u)$.

Now, we shortly discuss the relation to the configuration interaction (CI) approximation spaces known from quantum chemistry; see e.g. [170]. To this end, let us assume that we have the direct sum decompositions $U_\uparrow \oplus W_\uparrow = \mathcal{L}^2(\mathbb{R}^3)$ and $U_\downarrow \oplus W_\downarrow = \mathcal{L}^2(\mathbb{R}^3)$. Furthermore, let the finite-dimensional subspaces $\tilde{W}_\uparrow \subset W_\uparrow$ and $\tilde{W}_\downarrow \subset W_\downarrow$ be spanned by systems $\{\phi_p^\uparrow\}_{p \in \mathcal{N}_\uparrow'}$ and $\{\phi_p^\downarrow\}_{p \in \mathcal{N}_\downarrow'}$, respectively, where $\mathcal{N}_\uparrow' = \{1, \ldots, m_\uparrow\}$ and $\mathcal{N}_\downarrow' =$

4 Numerical Methods

$\{1, \ldots, m_\downarrow\}$. Then, if we set $\mathcal{B}_u$ equal to

$$\mathcal{B}_u^{CI} := \left\{ \mathfrak{A}^{(N,M_S)}\left(\vec{\mathbf{y}} \mapsto \prod_{q=1}^{n_\uparrow} \phi_{\nu_q}^\uparrow(\mathbf{y}_{p_q}) \prod_{q=n_\uparrow+1}^{|u|} \phi_{\nu_q}^\downarrow(\mathbf{y}_{p_q}) \prod_{p \in \mathcal{N} \setminus u} \psi_p(\mathbf{y}_p) \right) : \right.$$
$$\left. (\nu_1, \ldots, \nu_{|u|})^T \in \left((\mathcal{N}_\uparrow')^{n_\uparrow} \times (\mathcal{N}_\downarrow')^{|u|-n_\uparrow}\right) \cap \mathbb{Z}_u^{(N,M_S)} \right\} \subset \mathcal{B}_u^{(N,M_S)}, \quad (4.44)$$

the finite-dimensional approximation space $\tilde{\mathcal{V}}_{\{\gamma_u\}_u, \{\mathcal{B}_u^{CI}\}_u}^{(N,M_S)}$, i.e. the space given in (4.43) and (4.42) with $\{\phi_p^\uparrow\}_{p \in \mathcal{N}_\uparrow'}$, $\{\phi_p^\downarrow\}_{p \in \mathcal{N}_\downarrow'}$, corresponds to a restricted particle-wise subspace splitting as in (4.40) based on the splittings $U_\uparrow \oplus \tilde{W}_\uparrow$ and $U_\downarrow \oplus \tilde{W}_\downarrow$. In particular, a Galerkin approximation in $\tilde{\mathcal{V}}_{\{\gamma_u\}_u, \{\mathcal{B}_u^{CI}\}_u}^{(N,M_S)}$ of the lowest state as in (4.41) corresponds to a CI approximation. Note that in quantum chemistry it is common to additionally assume that $U_\uparrow \oplus W_\uparrow$ and $U_\downarrow \oplus W_\downarrow$ are orthogonal sums and also that the systems $\{\psi_p\}_{p \in \mathcal{N}_\uparrow'}$, $\{\phi_p^\uparrow\}_{p \in \mathcal{N}_\uparrow'}$, $\{\psi_p\}_{p \in \mathcal{N}_\downarrow'}$ and $\{\phi_p^\downarrow\}_{p \in \mathcal{N}_\downarrow'}$ are $\mathcal{L}^2$-orthonormal. Then, according to (2.94), the dimension of the subspace

$$\bigoplus_{u \subset \mathcal{N}, |u|=\mu} \tilde{\mathcal{W}}_{u,\mathcal{B}_u^{CI}}^{(N,M_S)}$$

is equal to

$$\sum_{0 \leq \mu_\uparrow \leq N_\uparrow, 0 \leq \mu_\downarrow \leq N_\downarrow, \mu_\uparrow + \mu_\downarrow = \mu} \binom{N_\uparrow}{\mu_\uparrow}\binom{m_\uparrow}{\mu_\uparrow}\binom{N_\downarrow}{\mu_\downarrow}\binom{m_\downarrow}{\mu_\downarrow}. \quad (4.45)$$

Hence, a CI approximation space associated with finite-order weights of order $q \leq N$ exhibits an upper bound of order $\mathcal{O}(N^q)$ for the number of degrees of freedom. Note that a finite-dimensional approximation space $\tilde{\mathcal{V}}_{\{\gamma_u\}_u, \{\mathcal{B}_u^{CI}\}_u}^{(N,M_S)}$ with finite-order weights $\{\gamma_u\}_{u \subset \mathcal{N}}$ of order $q = N$, i.e. $\gamma_u > 0$ for all $u \subset \mathcal{N}$, corresponds to the so-called full configuration interaction (FCI) space, which is spanned by

$$\mathcal{B}_{FCI}^{(N,M_S)} := \bigcup_{u \subset \mathcal{N}} \mathcal{B}_u^{CI}, \quad (4.46)$$

compare also (2.87). Here, according to (4.45) and Vandermonde's identity, we obtain the relation

$$|\mathcal{B}_{FCI}^{(N,M_S)}| = \binom{m_\uparrow + N_\uparrow}{N_\uparrow}\binom{m_\downarrow + N_\downarrow}{N_\downarrow} \leq \frac{(m_\uparrow + N_\uparrow)^{N_\uparrow}}{N_\uparrow!} \frac{(m_\downarrow + N_\downarrow)^{N_\downarrow}}{N_\downarrow!}$$

with regard to the cardinality number; compare (2.83).

Let us finally note that the construction scheme introduced in (4.43) and (4.42) allows for a more flexible choice of the finite-dimensional subspaces than in the framework of a common CI approximation corresponding to (4.44).

4.4 Adaptive scheme

In this section we present our heuristic approach for a choice of a sequence of finite-dimensional spaces

$$\left\{ V_\kappa := \tilde{\mathcal{V}}_{\{\gamma_u^{[\kappa]}\}_u, \{\mathcal{B}_u^{[\kappa]}\}_u}^{(N,M_S)} \subset \mathcal{V}_{\{\gamma_u^{[\kappa]}\}_u}^{(N,M_S)} \right\}_{\kappa \in \mathbb{N}_0} \quad (4.47)$$

4.4 Adaptive scheme

in the framework of a Galerkin approximation (3.19), where each approximation space $V_\kappa = \tilde{\mathcal{V}}^{(N,M_S)}_{\{\gamma_u^{[\kappa]}\}_u, \{\mathcal{B}_u^{[\kappa]}\}_u}$ corresponds to a subspace with finite-order weights as introduced in (4.43) with (4.42). Here, we first choose the initial approximation space

$$V_0 = \tilde{\mathcal{V}}^{(N,M_S)}_{\{\gamma_u^{[0]}\}_u, \{\mathcal{B}_u^{[0]}\}_u},$$

i.e. the finite-order weights $\{\gamma_u^{[0]}\}_u$ and initial subsets $\{\mathcal{B}_u^{[0]}\}_{u,\gamma_u>0}$, in an a priori fashion. Let us note that the initial approximation space is built with the help of one-particle subspaces (with just a few degrees of freedom) which provide an accurate representation of atomic orbitals, i.e. hydrogen-like wave functions. Then, starting from the initial subspace V_0, we apply a simple h-adaptive refinement scheme for the construction of an sequence of spanning systems $\{\mathcal{B}_u^{[\kappa]} \subset \mathcal{B}_u^{(N,M_S)}\}_{\kappa \in \mathbb{N}}$ in an a posteriori fashion. This sequence is iteratively constructed by the solution of the general linear eigenvalue problem and a refinement and expansion of the spanning system. The resulting sequence of approximation spaces

$$\left\{ V_\kappa = \tilde{\mathcal{V}}^{(N,M_S)}_{\{\gamma_u^{[\kappa]}\}_u, \{\mathcal{B}_u^{[\kappa]}\}_u} \right\}_{\kappa > 0}$$

hopefully gives us an efficient representation of the involved cusps. For a further reading an adaptive wavelet techniques; see e.g. [34, 41].

In the following we first discuss the a priori choice of the finite-order weights $\{\gamma_u^{[0]}\}_u$ and give the details for the construction of the initial sets $\{\mathcal{B}_u^{[0]}\}_{u,\gamma_u^{[0]}>0}$. Note here that we employ the second order multiscale Gaussian frame introduced in Section 4.2.2 as one-particle frames for the spaces $W_\uparrow$ and $W_\downarrow$. Finally, we present the technical details of our adaptive scheme to build the sequence of approximation spaces $\{V_\kappa\}_{\kappa>0}$.

4.4.1 A priori choice of the initial approximation space

Finite-order weights

In this thesis we do not consider an adaptive scheme with respect to the choice of weights and thus to the choice of subspaces which are taken into account in the restriction (4.40) of the subspace splitting (4.33). Instead, we a priori choose a fixed set of finite-order weights $\{\gamma_u\}_u$ and thereby the subspace $\mathcal{V}^{(N,M_S)}_{\{\gamma_u\}_u} \subset \mathcal{L}^2_{(N,M_S)}$. Here, we assume that the error with respect to an approximation in the a priori chosen space $\mathcal{V}^{(N,M_S)}_{\{\gamma_u\}_u}$ is negligibly small for such a choice of weights.

In particular, we employ sets of finite-order weights of order three so that we obtain a sum of subspaces according to (4.40) in the form

$$\mathcal{V}^{(N,M_S)}_{three} := \mathcal{V}^{zero} + \mathcal{V}^{one} + \mathcal{V}^{two\uparrow\downarrow} + \mathcal{V}^{two\uparrow\uparrow} + \mathcal{V}^{three} \subset \sum_{u \subset \mathcal{N}, |u| \leq 3} \mathcal{W}^{(N,M_S)}_u \subset \mathcal{L}^2_{(N,M_S)}, \quad (4.48)$$

where

$$\begin{aligned}
\mathcal{V}^{zero} &:= \mathcal{W}_{\emptyset}^{(N,M_S)}, \\
\mathcal{V}^{one} &:= \sum_{p_1 \in \mathcal{N}} \mathcal{W}_{\{p_1\}}^{(N,M_S)}, \\
\mathcal{V}^{two\uparrow\downarrow} &:= \sum_{p_1 \in \mathcal{N}_\uparrow} \sum_{p_2 \in \mathcal{N}_\downarrow} \mathcal{W}_{\{p_1,p_2\}}^{(N,M_S)}, \\
\mathcal{V}^{two\uparrow\uparrow} &:= \sum_{p_1,p_2 \in \mathcal{N}_\uparrow} \mathcal{W}_{\{p_1,p_2\}}^{(N,M_S)} + \sum_{p_1,p_2 \in \mathcal{N}_\downarrow} \mathcal{W}_{\{p_1,p_2\}}^{(N,M_S)}, \\
\mathcal{V}^{three} &:= \sum_{p_1,p_2 \in \mathcal{N}_\uparrow} \sum_{p_3 \in \mathcal{N}_\downarrow} \mathcal{W}_{\{p_1,p_2,p_3\}}^{(N,M_S)} + \sum_{p_1 \in \mathcal{N}_\uparrow} \sum_{p_2,p_3 \in \mathcal{N}_\downarrow} \mathcal{W}_{\{p_1,p_2,p_3\}}^{(N,M_S)}.
\end{aligned}$$

Here, $\mathcal{V}^{zero}$ is associated with the span of the rank-1 approximation according to (4.36). The space $\mathcal{V}^{one}$ is associated with a sum of spaces which is based on $\mathcal{V}^{zero}$. For each $1 \leq p \leq N$ the one-particle function ψ_p is replaced by an appropriate one-particle space; compare (4.34). Analogously, $\mathcal{V}^{two\uparrow\downarrow}$ is associated with the sum of spaces in which each pair of one-particle functions of opposite spin is replaced, $\mathcal{V}^{two\uparrow\uparrow}$ with the sum of spaces in which each pair of one-particle functions of same spin is replaced and $\mathcal{V}^{three}$ with the sum of spaces in which each triple of one-particle functions of not fully parallel spin is replaced.

Initial system

The spanning systems $\{\mathcal{B}_u^{(N,M_S)}\}_{u \subset \mathcal{N}, \gamma_u > 0}$ are given according to (4.42) for a set of finite-order weights $\{\gamma_u\}_u$. In particular, we employ a multiscale Gaussian frame as introduced in Section 4.2.2 as one-particle frames for $W_\uparrow$ and $W_\downarrow$. In the following we present our heuristic scheme for an a priori choice of the initial subsets $\mathcal{B}_u^{[0]} \subset \mathcal{B}_u^{(N,M_S)}$ for each $u \subset \mathcal{N}$ with $\gamma_u > 0$. Here, the resulting initial subspace

$$V_0 = \sum_{u \subset \mathcal{N}, \gamma_u > 0} \tilde{\mathcal{W}}_{u, \mathcal{B}_u^{[0]}}^{(N,M_S)} \subset \mathcal{V}_{\{\gamma_u^{[0]}\}_u}^{(N,M_S)}, \tag{4.49}$$

hopefully leads to systems $\mathcal{B}_u^{[\kappa]}$ for $\kappa > 0$, which allow for an efficient representation of cusps; compare Section 3.4.2.

First, we consider the one-particle case, i.e. the electronic Schrödinger equation for one electron

$$H = -\frac{1}{2}\Delta_\mathbf{x} - \frac{Z}{|\mathbf{x}|_2} \tag{4.50}$$

with an electronic Hamiltonian for a nucleus of atomic charge Z fixed at the origin. The corresponding eigenvalues read as $E^{(n)} = -\frac{Z^2}{2n^2}$ for $n = 1, 2, \ldots$ and the associated eigenfunctions are the well-known hydrogen-like wave functions. These wave functions can be written in form of a product of a spherical harmonic and a radial part. The radial part is composed of an exponential term and a polynomial term including Laguerre polynomials; see e.g. [137, 159]. For example, the eigenspace associated with the smallest eigenvalue $E^{(1)}$ is spanned by the $\mathcal{L}^2$-normalized s-type eigenfunction

$$\Psi_1^{(s)}(\mathbf{x}) = Z^{\frac{3}{2}} e^{(-Z|\mathbf{x}|_2)} \frac{1}{\sqrt{\pi}}.$$

4.4 Adaptive scheme

The eigenspace associated with the first excited eigenvalue $E^{[2]}$ is spanned by the $\mathcal{L}^2$-normalized s-type eigenfunction

$$\Psi_2^{(s)}(\mathbf{x}) = \frac{1}{4}\sqrt{\frac{2}{\pi}} Z^{\frac{3}{2}} \left(1 - |\mathbf{x}|_2 \frac{Z}{2}\right) e^{-\frac{1}{2}Z|\mathbf{x}|_2}$$

and the three $\mathcal{L}^2$-normalized p-type eigenfunctions

$$\Psi_2^{(p_d)}(\mathbf{x}) = \frac{1}{2\sqrt{2\pi}} Z^{\frac{5}{2}} x_d e^{-\frac{1}{2}Z|\mathbf{x}|_2}, \quad d = 1, 2, 3.$$

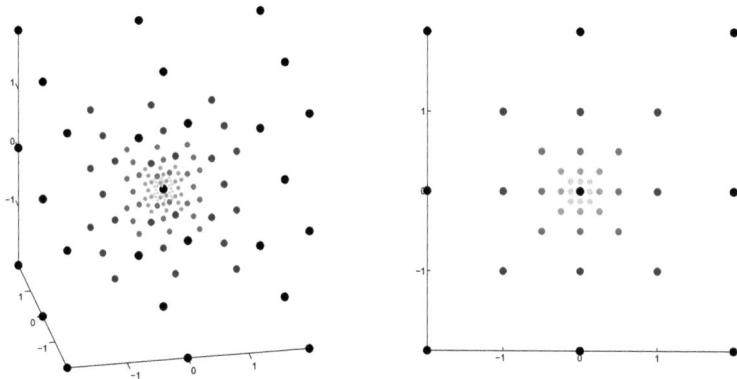

Figure 4.2: Localization peaks of basis functions in $\mathfrak{b}_{1,1/2}^{(5)}$ according to (4.51). Here, we use colors from black to light gray for the peaks associated with $\mathfrak{b}_{1,1/2}^{(1)}$, $\mathfrak{b}_{1,1/2}^{(2)} \setminus \mathfrak{b}_{1,1/2}^{(1)}$, $\mathfrak{b}_{1,1/2}^{(3)} \setminus \mathfrak{b}_{1,1/2}^{(2)}$, $\mathfrak{b}_{1,1/2}^{(4)} \setminus \mathfrak{b}_{1,1/2}^{(3)}$ and $\mathfrak{b}_{1,1/2}^{(5)} \setminus \mathfrak{b}_{1,1/2}^{(4)}$. On the right hand side we depict a slice plane, i.e. the (x_1, x_2)-plane, of the view on the left hand side.

Now, with the help of the second order multiscale Gaussian frame $\mathfrak{b}_{\sigma,c}$ according to (4.29) we introduce the initial system

$$\mathfrak{b}_{\sigma,c}^{(L)} := \{\varphi_{\sigma,c,0,\mathbf{j}} : |\mathbf{j}|_\infty \leq 1\} \cup \bigcup_{\mathbf{z} \in \mathcal{Z}} \left\{\psi_{\sigma,c,l,\mathbf{j}}^{[\mathbf{z}]} : l \leq L - 2, \mathbf{j} = -\mathbf{z}\right\} \subset \mathfrak{b}_{\sigma,c} \quad (4.51)$$

for $L \in \mathbb{N}$. We depict an example set in Figure 4.2.

For a better understanding, we employ the basis sets $\mathfrak{b}_{\sigma,c}^{(L)}$ for a Galerkin approximation of the smallest and the first excited eigenvalue of the hydrogen-like Hamiltonian (4.50). Some exemplarily computed eigenfunctions are depicted in Figure 4.3. In Figure 4.4 we show the approximation errors with respect to the exact eigenvalues

$$E^{(1)} = -Z^2/2 \quad \text{and} \quad E^{(1)} = -Z^2/8 \quad \text{for } Z \in \{1, 4\}.$$

4 Numerical Methods

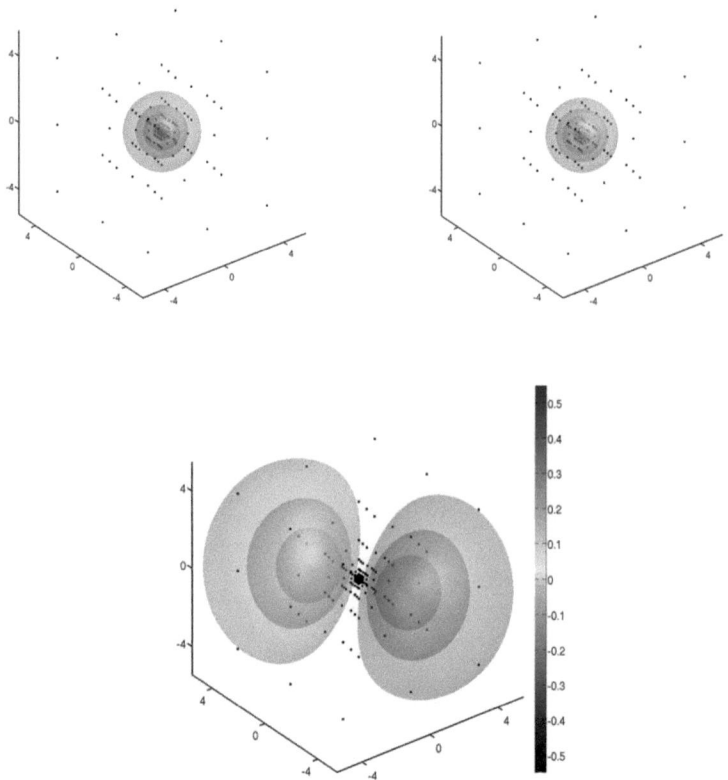

Figure 4.3: Visualization of different isosurfaces of approximately computed eigenfunctions of (4.50) with $Z = 1$ in initial subspaces $\text{span}(\mathfrak{b}^{(7)}_{1,1/4})$. Top: From left to right we have depicted an s-type eigenfunction associated with the smallest eigenvalue and an s-type eigenfunction associated with the first excited eigenvalue. Bottom: We have depicted a p-type eigenfunction associated with the first excited eigenvalue.

Our results indicate that the system $\mathfrak{b}^{(L)}_{\sigma,c}$ with appropriate chosen parameters σ, c and L depending on Z allows for an accurate approximation of the smallest and first excited eigenvalues and hence, for the approximation of certain electron-nuclei cusps.[12]

---

[12]Note that further numerical experiments indicate that a system $\mathfrak{b}^{(L)}_{\sigma,c}$ with appropriate chosen pa-

4.4 Adaptive scheme

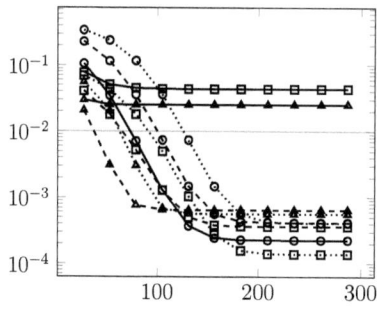
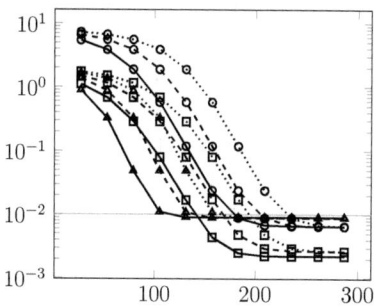

Figure 4.4: Error of the computed ground-state energy ($n = 1$) and the first excited state energy ($n = 2$) of a hydrogen-like atom according to (4.50) with $Z = 1$ (left) and $Z = 4$ (right) with respect to approximations in subspaces $\mathrm{span}(\mathfrak{b}_{1,c}^{(L)})$ for $L = 1, \ldots, 11$. We depict the error $|E^{(n)} - \tilde{E}_{(L)}^{(n)}|_2$ in hartree versus the number of degrees of freedom $|\mathfrak{b}_{1,c}^{(L)}|$. With regard to the scaling factor c, the line styles solid, dashed and dotted denote values $1/2$, $1/4$ and $1/8$, respectively. An error associated with a ground-state is denoted by a circle, an error associated with an s-type first excited state by a square and an error associated with a p-type first excited state by a triangle.

Let us now describe our heuristic scheme to a priori choose initial systems $\mathcal{B}_u^{[0]} \subset \mathcal{B}_u^{(N,M_S)}$ with regard to a Galerkin approximation in (4.47) of the minimal energy corresponding to a molecular Hamiltonian (3.1). Our scheme is based on the simple systems $\mathfrak{b}_{\sigma,c}^{(L)}$ according to (4.51). In this thesis we focus on the case of atomic and diatomic systems. Furthermore, we only apply approximation subspaces with finite-order weights of order three according to (4.48).

First, we consider an atomic system with N electrons and a nucleus of atomic charge Z_1 centered at $\mathbf{R}_1$, where we assume for reasons of simplicity $\mathbf{R}_1 = \mathbf{0}$. Let the parameters σ, c and L be fixed. Then, in the case of spanning systems associated with finite-order weights $\gamma_u > 0$ with $|u| = 1$, we set

$$\mathcal{B}_{\{p_1\}}^{[0]} = \mathcal{B}_{\{p_1\}}^{(N,M_S)} \cap \left\{ \mathfrak{A}^{(N,M_S)}\left(\vec{\mathbf{y}} \mapsto \phi(\mathbf{y}_{p_1}) \prod_{q \in \mathcal{N} \setminus \{p_1\}} \psi_q(\mathbf{y}_q)\right) : \phi \in \mathfrak{b}_{\sigma,c}^{(L)} \right\} \quad (4.52)$$

for $p_1 \in \mathcal{N}$. For spanning systems according to finite-order weights $\gamma_u > 0$, $|u| \geq 2$ the idea is to employ certain tensor product functions with localization peaks at or close to the electronic cups. In this way, we set

$$\mathcal{B}_{\{p_1,p_2\}}^{[0]} = \mathcal{B}_{\{p_1,p_2\}}^{(N,M_S)} \cap \left\{ \mathfrak{A}^{(N,M_S)}\left(\vec{\mathbf{y}} \mapsto \phi(\mathbf{y}_{p_1})\phi(\mathbf{y}_{p_2}) \prod_{q \in \mathcal{N} \setminus \{p_1,p_2\}} \psi_q(\mathbf{y}_q)\right) : \phi \in \mathfrak{b}_{\sigma,c}^{(L)} \right\}, \quad (4.53)$$

rameters σ, c and L depending on Z allows also for an accurate approximation of higher excited eigenvalues. However, in this thesis we focus on the case of the two smallest eigenvalues since we are mainly interested in the ground-state and first-exited states of small molecular systems.

for $p_1 \in \mathcal{N}_\uparrow$, $p_2 \in \mathcal{N}_\downarrow$. In the case of two electrons of parallel spin, i.e. $p_1, p_2 \in \mathcal{N}_\uparrow$ or $p_1, p_2 \in \mathcal{N}_\downarrow$, the set $\mathcal{B}^{[0]}_{\{p_1,p_2\}}$ according to (4.53) is empty due to the partially antisymmetry condition. Hence, in this case we employ a modified variant of the set defined in (4.53). To this end, we first introduce a mapping which gives the unique center point (localization peak) of a one-particle function of the frame $\mathfrak{b}_{\sigma,c}$, i.e.

$$\mathbf{C}: \mathfrak{b}_{\sigma,c} \to \mathbb{R}^3 : \phi \mapsto \begin{cases} (c)^{-1}\mathbf{j} & \text{for } \phi = \varphi_{\sigma,c,0,\mathbf{j}}, \\ (c2^l)^{-1}(\mathbf{j} - \tfrac{1}{2}\mathbf{z}) & \text{for } \phi = \psi^{[\mathbf{z}]}_{\sigma,c,l,\mathbf{j}}. \end{cases}$$

Then, for $\phi \in \mathfrak{b}^{(L)}_{\sigma,c}$ we introduce a specific set of nearest neighbor one-particle functions of the frame $\mathfrak{b}_{\sigma,c}$ by

$$\operatorname{Ng}(\phi) := \begin{cases} \{\tilde{\phi} \in \mathfrak{b}_{\sigma,c} : \mathbf{C}(\tilde{\phi}) = \mathbf{C}(\phi) + c^{-1}\mathbf{z}', \mathbf{z}' \in \mathcal{Z}'\} & \text{for } \phi \in \{\varphi_{\sigma,c,0,\mathbf{j}}\}_{\mathbf{j} \in \mathbb{Z}^3}, \\ \{\tilde{\phi} \in \mathfrak{b}_{\sigma,c} : \mathbf{C}(\tilde{\phi}) = \mathbf{C}(\phi) + (c2^{l+1})^{-1}\mathbf{z}', \mathbf{z}' \in \mathcal{Z}'\} & \text{for } \phi \in \{\psi^{[\mathbf{z}]}_{\sigma,c,l,\mathbf{j}}\}_{\mathbf{j} \in \mathbb{Z}^3, \mathbf{z} \in \mathcal{Z}}, \end{cases} \quad (4.54)$$

where $\mathcal{Z}' := \{\mathbf{z}' \in \{-1, 0, 1\}^3 : |\mathbf{z}|_1 \leq 1\}$. Let us remark that there are many other possibilities to define an appropriate set $\operatorname{Ng}(\phi)$ of neighboring one-particle functions of ϕ. Now, for either $p_1, p_2 \in \mathcal{N}_\uparrow$ or $p_1, p_2 \in \mathcal{N}_\downarrow$ we set

$$\mathcal{B}^{[0]}_{\{p_1,p_2\}} = \mathcal{B}^{(N,M_S)}_{\{p_1,p_2\}} \cap$$
$$\left(\left\{\mathfrak{A}^{(N,M_S)}\left(\vec{\mathbf{y}} \mapsto \tilde{\phi}(\mathbf{y}_{p_1})\phi(\mathbf{y}_{p_2}) \prod_{q \in \mathcal{N} \setminus \{p_1,p_2\}} \psi_q(\mathbf{y}_q)\right) : \phi \in \mathfrak{b}^{(L)}_{\sigma,c}, \tilde{\phi} \in \operatorname{Ng}(\phi)\right\} \right. \quad (4.55)$$
$$\left. \cup \left\{\mathfrak{A}^{(N,M_S)}\left(\vec{\mathbf{y}} \mapsto \phi(\mathbf{y}_{p_1})\tilde{\phi}(\mathbf{y}_{p_2}) \prod_{q \in \mathcal{N} \setminus \{p_1,p_2\}} \psi_q(\mathbf{y}_q)\right) : \phi \in \mathfrak{b}^{(L)}_{\sigma,c}, \tilde{\phi} \in \operatorname{Ng}(\phi)\right\}\right).$$

In the three electron case, i.e. for either $p_1, p_2 \in \mathcal{N}_\uparrow, p_3 \in \mathcal{N}_\uparrow$ or $p_1, p_2 \in \mathcal{N}_\downarrow, p_3 \in \mathcal{N}_\uparrow$, we set

$$\mathcal{B}^{[0]}_{\{p_1,p_2,p_3\}} = \mathcal{B}^{(N,M_S)}_{\{p_1,p_2,p_3\}} \cap$$
$$\left(\left\{\mathfrak{A}^{(N,M_S)}\left(\vec{\mathbf{y}} \mapsto \tilde{\phi}(\mathbf{y}_{p_1})\phi(\mathbf{y}_{p_2})\phi(\mathbf{y}_{p_3}) \prod_{q \in \mathcal{N} \setminus \{p_1,p_2,p_3\}} \psi_q(\mathbf{y}_q)\right) : \phi \in \mathfrak{b}^{(2)}_{\sigma,c}, \tilde{\phi} \in \operatorname{Ng}(\phi)\right\}\right.$$
$$\left.\cup \left\{\mathfrak{A}^{(N,M_S)}\left(\vec{\mathbf{y}} \mapsto \phi(\mathbf{y}_{p_1})\tilde{\phi}(\mathbf{y}_{p_2})\phi(\mathbf{y}_{p_3}) \prod_{q \in \mathcal{N} \setminus \{p_1,p_2,p_3\}} \psi_q(\mathbf{y}_q)\right) : \phi \in \mathfrak{b}^{(2)}_{\sigma,c}, \tilde{\phi} \in \operatorname{Ng}(\phi)\right\}\right)$$
$$(4.56)$$

independent of the value of the parameter L.

In the case of a diatomic systems with nuclei of atomic charge Z_1 and Z_2 centered at $\mathbf{R}_1$ and $\mathbf{R}_2$, we assume $\mathbf{R}_1 = \mathbf{0}$, $\mathbf{R}_2 = (R_{12}, 0, 0)^T$, $R_{12} > 0$ and choose for the scaling parameter c a value of $c = 2/R_{12}$. Let us further assume that the set of one-particle functions $\{\psi_p\}_{p \in \mathcal{N}}$, with $U_\uparrow = \operatorname{span}(\{\psi_p\}_{p \in \mathcal{N}_\uparrow})$ and $U_\downarrow = \operatorname{span}(\{\psi_p\}_{p \in \mathcal{N}_\downarrow})$, can be partitioned into three distinct sets. First, a set $\{\psi_p\}_{p \in \mathcal{R}_1}$, $\mathcal{R}_1 \subset \mathcal{N}$, associated with $\mathbf{R}_1$. Second, a set $\{\psi_p\}_{p \in \mathcal{R}_2}$, $\mathcal{R}_2 \subset \mathcal{N}$, associated with $\mathbf{R}_2$. And finally, a set

$\{\psi_p\}_{p \in \mathcal{R}_{1,2}}$, $\mathcal{R}_{1,2} \subset \mathcal{N}$, associated with both, $\mathbf{R}_1$ and $\mathbf{R}_2$. Moreover, we denote by $\mathfrak{b}_{\sigma,c,\mathbf{R}}^{(L)}$ the set of functions according to (4.51) which are centered at $\mathbf{R}$ instead of the origin, i.e. $\mathfrak{b}_{\sigma,c,\mathbf{R}}^{(L)} := \{\phi(\mathbf{x}-\mathbf{R}) : \phi \in \mathfrak{b}_{\sigma,c}^{(L)}\}$. Note that with the assumptions $\mathbf{R}_1 = \mathbf{0}$ and $c = 2/R_{12}$ the inclusions $\mathfrak{b}_{\sigma,c,\mathbf{R}_1}^{(L)}, \mathfrak{b}_{\sigma,c,\mathbf{R}_2}^{(L)}, \mathfrak{b}_{\sigma,c,(\mathbf{R}_1+\mathbf{R}_2)/2}^{(L)} \subset \mathfrak{b}_{\sigma,c}$ in particular hold. Moreover, to shorten notation we introduce the subsets $\mathfrak{b}_{\sigma,c,p}^{(L)}$ of $\mathfrak{b}_{\sigma,c}$ for $p \in \mathcal{N}$ by

$$\mathfrak{b}_{\sigma,c,p}^{(L)} := \begin{cases} \mathfrak{b}_{\sigma,c,\mathbf{R}_1}^{(L)} & \text{for } p \in \mathcal{R}_1, \\ \mathfrak{b}_{\sigma,c,\mathbf{R}_2}^{(L)} & \text{for } p \in \mathcal{R}_2, \\ \mathfrak{b}_{\sigma,c,(\mathbf{R}_1+\mathbf{R}_2)/2}^{(L)} & \text{for } p \in \mathcal{R}_{1,2}. \end{cases}$$

Now, analogous to (4.52), we set

$$\mathcal{B}_{\{p_1\}}^{[0]} = \mathcal{B}_{\{p_1\}}^{(N,M_S)} \cap \left\{ \mathfrak{A}^{(N,M_S)}\left(\vec{\mathbf{y}} \mapsto \phi(\mathbf{y}_{p_1}) \prod_{q \in \mathcal{N} \setminus \{p_1\}} \psi_q(\mathbf{y}_q)\right) : \phi \in \mathfrak{b}_{\sigma,c,p_1}^{(L)} \right\} \quad (4.57)$$

for $p_1 \in \mathcal{N}$. Analogous to (4.53), we set

$$\mathcal{B}_{\{p_1,p_2\}}^{[0]} = \mathcal{B}_{\{p_1,p_2\}}^{(N,M_S)} \cap$$
$$\left\{ \mathfrak{A}^{(N,M_S)}\left(\vec{\mathbf{y}} \mapsto \phi(\mathbf{y}_{p_1})\phi(\mathbf{y}_{p_2}) \prod_{q \in \mathcal{N} \setminus \{p_1,p_2\}} \psi_q(\mathbf{y}_q)\right) : \phi \in \mathfrak{b}_{\sigma,c,p_1}^{(L)} \cap \mathfrak{b}_{\sigma,c,p_2}^{(L)} \right\} \quad (4.58)$$

for $p_1 \in \mathcal{N}_\uparrow$ and $p_2 \in \mathcal{N}_\downarrow$. Analogous to (4.55), we set

$$\mathcal{B}_{\{p_1,p_2\}}^{[0]} = \mathcal{B}_{\{p_1,p_2\}}^{(N,M_S)} \cap$$
$$\left(\left\{ \mathfrak{A}^{(N,M_S)}\left(\vec{\mathbf{y}} \mapsto \tilde{\phi}(\mathbf{y}_{p_1})\phi(\mathbf{y}_{p_2}) \prod_{q \in \mathcal{N} \setminus \{p_1,p_2\}} \psi_q(\mathbf{y}_q)\right) \right.\right.$$
$$\left. : \phi \in \mathfrak{b}_{\sigma,c,p_1}^{(L)} \cap \mathfrak{b}_{\sigma,c,p_2}^{(L)}, \tilde{\phi} \in \mathrm{Ng}(\phi) \right\}$$
$$\cup \left\{ \mathfrak{A}^{(N,M_S)}\left(\vec{\mathbf{y}} \mapsto \phi(\mathbf{y}_{p_1})\tilde{\phi}(\mathbf{y}_{p_2}) \prod_{q \in \mathcal{N} \setminus \{p_1,p_2\}} \psi_q(\mathbf{y}_q)\right) \right.$$
$$\left.\left. : \phi \in \mathfrak{b}_{\sigma,c,p_1}^{(L)} \cap \mathfrak{b}_{\sigma,c,p_2}^{(L)}, \tilde{\phi} \in \mathrm{Ng}(\phi) \right\}\right) \quad (4.59)$$

for either $p_1, p_2 \in \mathcal{N}_\uparrow$ or $p_1, p_2 \in \mathcal{N}_\downarrow$. And analogous to (4.56), we set

$$\mathcal{B}^{[0]}_{\{p_1,p_2,p_2\}} = \mathcal{B}^{(N,M_S)}_{\{p_1,p_2,p_3\}} \cap$$

$$\left(\left\{ \mathfrak{A}^{(N,M_S)} \left(\vec{y} \mapsto \tilde{\phi}(\mathbf{y}_{p_1}) \phi(\mathbf{y}_{p_2}) \phi(\mathbf{y}_{p_3}) \prod_{q \in \mathcal{N} \setminus \{p_1,p_2,p_3\}} \psi_q(\mathbf{y}_q) \right) \right. \right.$$
$$: \phi \in \mathfrak{b}^{(2)}_{\sigma,c,p_1} \cap \mathfrak{b}^{(2)}_{\sigma,c,p_2} \cap \mathfrak{b}^{(2)}_{\sigma,c,p_3}, \tilde{\phi} \in \mathrm{Ng}(\phi) \Big\}$$
$$\cup \left\{ \mathfrak{A}^{(N,M_S)} \left(\vec{y} \mapsto \phi(\mathbf{y}_{p_1}) \tilde{\phi}(\mathbf{y}_{p_2}) \phi(\mathbf{y}_{p_3}) \prod_{q \in \mathcal{N} \setminus \{p_1,p_2,p_3\}} \psi_q(\mathbf{y}_q) \right) \right.$$
$$\left. \left. : \phi \in \mathfrak{b}^{(2)}_{\sigma,c,p_1} \cap \mathfrak{b}^{(2)}_{\sigma,c,p_2} \cap \mathfrak{b}^{(2)}_{\sigma,c,p_3}, \tilde{\phi} \in \mathrm{Ng}(\phi) \right\} \right) \quad (4.60)$$

for either $p_1, p_2 \in \mathcal{N}_\uparrow$, $p_3 \in \mathcal{N}_\uparrow$ or $p_1, p_2 \in \mathcal{N}_\downarrow$, $p_3 \in \mathcal{N}_\uparrow$. Note that in the atomic case the sets defined by (4.57), (4.58), (4.59) and (4.60) coincide with (4.52), (4.53), (4.55) and (4.56), respectively.

Altogether, according to (4.49) and (4.48), we set the initial subspace $V_0 \subset \mathcal{V}^{(N,M_S)}_{three}$ equal to the span of the following initial system

$$\mathcal{B}^{[0]}_{three} := \left\{ \mathfrak{A}^{(N,M_S)} \left(\vec{y} \mapsto \prod_{q \in \mathcal{N}} \psi_q(\mathbf{y}_q) \right) \right\} \cup \bigcup_{p_1 \in \mathcal{N}} \mathcal{B}^{[0]}_{\{p_1\}} \cup \bigcup_{p_1,p_2 \in \mathcal{N}} \mathcal{B}^{[0]}_{\{p_1,p_2\}}$$
$$\cup \bigcup_{p_1,p_2 \in \mathcal{N}_\uparrow} \bigcup_{p_3 \in \mathcal{N}_\downarrow} \mathcal{B}^{[0]}_{\{p_1,p_2,p_3\}} \cup \bigcup_{p_1,p_2 \in \mathcal{N}_\downarrow} \bigcup_{p_3 \in \mathcal{N}_\uparrow} \mathcal{B}^{[0]}_{\{p_1,p_2,p_3\}}. \quad (4.61)$$

4.4.2 A posteriori choice of the sequence of approximation spaces

In the following we describe our adaptive refinement and expansion scheme to choose a sequence of finite-dimensional spaces according to (4.47) in the framework of a Galerkin approximation, where we focus on the lowest state with respect to (N, M_S) only.

We employ an a priori chosen initial system from (4.61). However, if we consider molecular systems with more than three electrons, we additionally switch on the subspace associated with the weight $\gamma_\mathcal{N}$ in (4.40) and (4.49). Hence, for $N > 3$ we enrich the initial

4.4 Adaptive scheme

system given in (4.61) by the partially antisymmetric functions

$$\mathcal{B}_{other}^{[0]} := \mathcal{B}_{\mathcal{N}}^{(N,M_S)} \cap \left(\left\{ \mathfrak{A}^{(N,M_S)} \left(\vec{y} \mapsto \varphi_{c,\sigma,0,0}(\mathbf{y}_1) \prod_{p=2}^{N_\uparrow} \psi_{c,\sigma,p-2,0}^{[1]}(\mathbf{y}_p) \right. \right. \right.$$
$$\left. \times \varphi_{c,\sigma,0,0}(\mathbf{y}_{N_\uparrow+1}) \prod_{p=N_\uparrow+2}^{N} \psi_{c,\sigma,p-N_\uparrow-2,0}^{[1]}(\mathbf{y}_p) \right) \right\}$$
$$\cup \left\{ \mathfrak{A}^{(N,M_S)} \left(\vec{y} \mapsto \varphi_{c,\sigma,0,0}(\mathbf{y}_1) \prod_{p=2}^{N_\uparrow} \psi_{c,\sigma,p-2,-1}^{[1]}(\mathbf{y}_p) \right. \right.$$
$$\left. \left. \left. \times \varphi_{c,\sigma,0,0}(\mathbf{y}_{N_\uparrow+1}) \prod_{p=N_\uparrow+2}^{N} \psi_{c,\sigma,p-N_\uparrow-2,-1}^{[1]}(\mathbf{y}_p) \right) \right\} \right)$$
(4.62)

in the atomic case, where $\mathbf{1} = (1,1,1)^T$. Analogously, in the diatomic case we enrich the initial system given in (4.61) by functions in (4.62) translated to $\mathbf{R}_1$, $\mathbf{R}_2$ and $(\mathbf{R}_1 + \mathbf{R}_2)/2$, respectively.

Furthermore, we apply a coarsening step on the a priori chosen system $\mathcal{B}_{three}^{[0]} \cup \mathcal{B}_{other}^{[0]}$. Let us discuss this coarsening step in the following. Let $\mathcal{B} = \{\Phi_\mu\}_{\mu \in \{1,...,M\}}$ a $\mathcal{L}^2$-normalized basis set of a finite-dimensional subspace of $\mathcal{L}^2_{(N,M_S)}$. We denote the lowest state and its energy associated with a subspace spanned by a subset $\mathcal{B}' \subset \mathcal{B}$ by

$$\tilde{\Psi}^{(1)}(\mathcal{B}') = \underset{\Psi \in \text{span}(\mathcal{B}'), \|\Psi\|=1}{\arg \min} \langle \Psi, H\Psi \rangle, \quad \tilde{E}^{(1)}(\mathcal{B}') = \underset{\Psi \in \text{span}(\mathcal{B}'), \|\Psi\|=1}{\min} \langle \Psi, H\Psi \rangle,$$

respectively. Let us further assume that with respect to the basis set $\mathcal{B} = \{\Phi_\mu\}_{\mu \in \{1,...,M\}}$ the absolute values of the coefficients $|v_1|, \ldots, |v_{|\mathcal{B}|}|$ associated with the representation of the lowest state $\tilde{\Psi}^{(1)}(\mathcal{B}) = \sum_{\mu=1}^{|\mathcal{B}|} v_\mu^{\mathcal{B}} \Phi_\mu$ are unique. Then, for $\delta > 0$ we introduce a coarsened system

$$\text{coarse}_\delta(\mathcal{B}) := \underset{\{\mathcal{B}' \subset \mathcal{B} : |\tilde{E}^{(1)}(\mathcal{B}') - \tilde{E}^{(1)}(\mathcal{B})| \leq \delta, \mathcal{B}' = \mathcal{B}(\nu), \nu \in \mathbb{N}_0\}}{\arg \min} |\mathcal{B}'|,$$

where

$$\mathcal{B}(\nu) := \{\Phi_\mu \in \mathcal{B} : |v_\mu^{\mathcal{B}}| \geq 2^{-\nu}\}.$$

In the following we fix δ to a value of $1/100$ kcal/mol $\approx 1.59_{-5}$ hartree and set the resulting initial system equal to

$$\mathcal{B}^{[0]} := \text{coarse}_\delta \left(\mathcal{B}_{three}^{[0]} \cup \mathcal{B}_{other}^{[0]} \right).$$
(4.63)

We denote its associated finite-dimensional initial subspace by

$$V_0 = \text{span}(\mathcal{B}^{[0]}).$$

To describe our adaptive scheme, we introduce another set of neighbors by

$$\text{Ng}^+(\phi) :=$$
$$\begin{cases} \{\tilde{\phi} \in \mathfrak{b}_{\sigma,c} : \mathbf{C}(\tilde{\phi}) = \mathbf{C}(\phi) + (c2)^{-1}\mathbf{z}', \mathbf{z}' \in \mathcal{Z}'\} & \text{for } \phi \in \{\varphi_{\sigma,c,0,\mathbf{j}}\}_{\mathbf{j} \in \mathbb{Z}^3}, \\ \{\tilde{\phi} \in \mathfrak{b}_{\sigma,c} : \mathbf{C}(\tilde{\phi}) = \mathbf{C}(\phi) + (c2^{l+2})^{-1}\mathbf{z}', \mathbf{z}' \in \mathcal{Z}'\} & \text{for } \phi \in \{\psi_{\sigma,c,l,\mathbf{j}}^{[\mathbf{z}]}\}_{\mathbf{j} \in \mathbb{Z}^3, \mathbf{z} \in \mathcal{Z}}, \end{cases}$$
(4.64)

4 Numerical Methods

besides the set of neighbors Ng(ϕ) for a one-particle function $\phi \in \mathfrak{b}_{\sigma,c}$ as in (4.54). Furthermore, we generalize the definition of the set of neighbors Ng and Ng$^+$ to the case of a partially antisymmetric N-particle function Φ in the system $\mathcal{B}_u^{(N,M_S)}$ according to (4.42). Let $\Phi = \mathfrak{A}^{(N,M_S)}\left(\vec{y} \mapsto \prod_{q \in u} \phi_q(\mathbf{y}_q) \prod_{p \in \mathcal{N} \setminus u} \psi_p(\mathbf{y}_p)\right)$ be in $\mathcal{B}_u^{(N,M_S)}$, where $\phi_q \in \mathfrak{b}_{c,\sigma}$ for $q \in u$. Then, we define the set

$$\mathrm{Ng}(\Phi) := \bigcup_{q' \in u} \left\{ \mathfrak{A}^{(N,M_S)}\left(\vec{y} \mapsto \tilde{\phi}_{q'}(\mathbf{y}_{q'}) \prod_{q \in u \setminus \{q'\}} \phi_q(\mathbf{y}_q) \prod_{p \in \mathcal{N} \setminus u} \psi_p(\mathbf{y}_p)\right) \in \mathcal{B}_u^{(N,M_S)} \right.$$
$$\left. : \tilde{\phi}_{q'} \in \mathrm{Ng}(\phi_{q'}) \right\}$$

and the set

$$\mathrm{Ng}^+(\Phi) := \bigcup_{q' \in u} \left\{ \mathfrak{A}^{(N,M_S)}\left(\vec{y} \mapsto \tilde{\phi}_{q'}(\mathbf{y}_{q'}) \prod_{q \in u \setminus \{q'\}} \phi_q(\mathbf{y}_q) \prod_{p \in \mathcal{N} \setminus u} \psi_p(\mathbf{y}_p)\right) \in \mathcal{B}_u^{(N,M_S)} \right.$$
$$\left. : \tilde{\phi}_{q'} \in \mathrm{Ng}^+(\phi_{q'}) \right\}.$$

Now, we introduce the adaptive refinement and expansion scheme[13] which we use in this thesis by Algorithm 4.1. In this algorithm $\zeta_\mu^{\mathcal{B},S_\mathcal{B}} : \mathbb{R}^M \to \mathbb{R}_+$ denotes an appropriate function to (heuristically) estimate the importance of the basis function Φ_μ in an $\mathcal{L}^2$-normalized basis set $\mathcal{B} = \{\Phi_1, \ldots, \Phi_M\}$. We write it in the form

$$\zeta_\mu^{\mathcal{B},S_\mathcal{B}} : \mathbb{R}^M \to \mathbb{R}_+ : (v_1^\mathcal{B}, \ldots, v_M^\mathcal{B})^T \mapsto \zeta_\mu^\mathcal{B}\left(S_\mathcal{B}(v_1^\mathcal{B}, \ldots, v_M^\mathcal{B})^T\right), \quad (4.66)$$

where $S_\mathcal{B} \in \mathbb{R}^{M \times M}$ denotes an appropriate non-singular transformation matrix and $\zeta^\mathcal{B}$ is the function which just takes the absolute value of the corresponding coefficient into account, i.e.

$$\zeta_\mu^\mathcal{B} : \mathbb{R}^M \to \mathbb{R}_+ : (v_1^\mathcal{B}, \ldots, v_M^\mathcal{B})^T \mapsto |v_\mu^\mathcal{B}|.$$

Let us now consider the simple test case of a hydrogen-like atom according to (4.50), where we apply span($\mathfrak{b}_{1,c}^{(L)}$) according to (4.51) as initial subspace and set $S_\mathcal{B}$ to the respective identity matrix. In Figure 4.5 we give the numerical results for $L = 7$. In particular, our results show no saturation behaviour for the considered sequences of subspaces as compared with our results presented in Figure 4.4 with respect to the initial subspaces. Note that the measurement of approximation rates by a log-log fit results in approximation rates with respect to the number of degrees of freedom of orders in the range between approximately $-3/2$ and -2.[14]

Let us now discuss the choice of the transformation matrix for $\zeta_\mu^{\mathcal{B},S_\mathcal{B}}$ in the framework of the particle-wise subspace splitting. Here, we apply a transformation matrix $S_\mathcal{B} \in \mathbb{R}^{M \times M}$

---

[13] Note that the set of neighbors Ng(ϕ) can be seen as related to an *expansion* of a finite basis $\mathfrak{b} \subset \mathfrak{b}_{\sigma,c}$, $\phi \in \mathfrak{b}$. Similar the set of neighbors Ng$^+(\phi)$ can be seen as related to a *refinement* of this finite basis set.

[14] Note that due to Theorem 3.2 and relation (3.20), an order of approximately $-2/3$ with respect to the number of degrees of freedom can be expected in the case of a linear approximation scheme.

4.4 Adaptive scheme

Algorithm 4.1 Expansion and refinement scheme for chosen $0 < \varepsilon < 1$ and $\kappa_{\max} \geq 1$.

1. Set $\kappa = 1$ and $\mathcal{B} = \mathcal{B}^{[0]} = \text{coarse}_\delta(\mathcal{B}^{[0]}_{three} \cup \mathcal{B}^{[0]}_{other})$.

2. If $\kappa > \kappa_{\max}$ stop.

3. Solve
$$\tilde{\Psi}^{(1)} = \sum_{\mu=1}^{|\mathcal{B}|} v^{\mathcal{B}}_\mu \Phi_\mu = \operatorname*{arg\,min}_{\Psi \in \text{span}(\mathcal{B}), \|\Psi\|=1} \langle \Psi, H\Psi \rangle,$$
$$\tilde{E}^{(1)} = \min_{\Psi \in \text{span}(\mathcal{B}), \|\Psi\|=1} \langle \Psi, H\Psi \rangle. \tag{4.65}$$

4.   a) Set $\mathcal{B}' = \left\{\Phi_\mu \in \mathcal{B} : \zeta^{\mathcal{B},S_\mathcal{B}}_\mu\left((v^{\mathcal{B}}_1, \ldots, v^{\mathcal{B}}_{|\mathcal{B}|})^T\right) \geq \varepsilon\right\}$ and $\mathcal{B}^{\text{old}} = \mathcal{B}$.
     b) Set $\mathcal{B} = \{\Phi' : \Phi' \in \text{Ng}(\Phi), \Phi \in \mathcal{B}'\} \cup \{\Phi' : \Phi' \in \text{Ng}^+(\Phi), \Phi \in \mathcal{B}'\}$.
     c) If $\mathcal{B} \setminus \mathcal{B}^{\text{old}} \neq \emptyset$ then go to step 3.

5. Set $\tilde{E}^{(1)}_\kappa = \tilde{E}^{(1)}$ and $\tilde{\Psi}^{(1)}_\kappa = \tilde{\Psi}^{(1)}$. Also, we set $\mathcal{B}^{[\kappa]} = \mathcal{B}$ and denote the associated subspace by $V_\kappa = \text{span}(\mathcal{B}^{[\kappa]})$ and its number of degrees of freedom by $M_\kappa = |\mathcal{B}^{[\kappa]}|$.

6. Set $\kappa \leftarrow \kappa + 1$ and go to step 2.

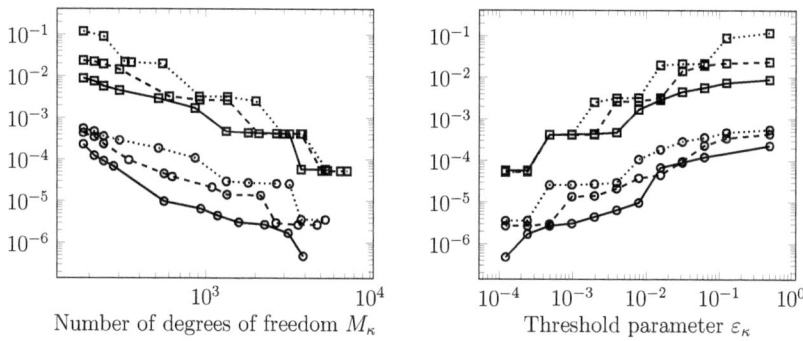

Figure 4.5: Approximation error of a hydrogen-like atom according to (4.50) in a sequence of subspaces created by Algorithm 4.1 starting with a subspace $\mathcal{B} = \text{span}(\mathfrak{b}^{(7)}_{1,c})$. We depict the error $|E^{(1)} - \tilde{E}^{(1)}_\kappa|_2$ in hartree versus the number of degrees of freedom M_κ (left) and versus the threshold parameter $\varepsilon_\kappa = 2^{-\kappa}$ (right). With regard to the scaling factor c, the line styles solid, dashed and dotted denote a value of $1/2$, $1/4$ and $1/8$, respectively. An error associated with $Z = 1$ is denoted by a circle and an error associated with $Z = 4$ is denoted by a square.

which results from a partial orthogonalization of a basis $\mathcal{B}$. This is discussed in more detail in the following.

For $m_1 + m_2 = M$ we write the positive definite symmetric mass matrix $B \in \mathbb{R}^{M \times M}$ according to (4.2) in the form of a partitioned matrix

$$B = \begin{pmatrix} B_{11} & B_{12} \\ B_{12}^T & B_{22} \end{pmatrix}$$

with submatrices $B_{11} \in \mathbb{R}^{m_1 \times m_1}$, $B_{12} \in \mathbb{R}^{m_1 \times m_2}$ and $B_{22} \in \mathbb{R}^{m_1 \times m_2}$. Furthermore, we introduce an $M \times M$ transformation matrix in block structure by

$$P_O := \begin{pmatrix} I_{11} & -B_{11}^{-1} B_{12} \\ 0 & I_{22} \end{pmatrix},$$

where $I_{11} \in \mathbb{R}^{m_1 \times m_1}$ and $I_{22} \in \mathbb{R}^{m_2 \times m_2}$ denote the respective identity matrices. Then, we can transform the basis set to a partially orthogonal basis set, i.e. we can write the mass matrix in block diagonal form

$$P_O^T \begin{pmatrix} B_{11} & B_{12} \\ B_{12}^T & B_{22} \end{pmatrix} P_O = \begin{pmatrix} B_{11} & 0 \\ 0 & B_{22} - B_{12}^T B_{11}^{-1} B_{12} \end{pmatrix}. \tag{4.67}$$

Note that the term $B_{22} - B_{12}^T B_{11}^{-1} B_{12}$ corresponds to the Schur complement of B_{11} in B [197]. Furthermore, we define the matrix $P_N := \mathrm{diag}(P_O^T B P_O)^{-\frac{1}{2}}$ for normalization and introduce an $M \times M$ transformation matrix by

$$S := (P_O P_N)^{-1} = \mathrm{diag}(P_O^T B P_O)^{\frac{1}{2}} P_O^{-1} \tag{4.68}$$

for the $\mathcal{L}^2$-normalized basis set $\mathcal{B}$. Here, P_O^{-1} can be easily computed in terms of submatrices since it is given by

$$P_O^{-1} = \begin{pmatrix} I_{11} & B_{11}^{-1} B_{12} \\ 0 & I_{22} \end{pmatrix}.$$

Now, let the pair $\tilde{E} \in \mathbb{R}$, $v \in \mathbb{R}^M$ solve the general eigenvalue problem in matrix form (4.1) resulting from a Galerkin approximation in the $\mathcal{L}^2$-normalized basis set $\mathcal{B} = \{\Phi_1, \ldots, \Phi_M\}$. Then, the pair $\tilde{E}$, $v' = Sv$ solves the general eigenvalue problem in matrix form given by

$$S^{-T} A S^{-1} v' = \tilde{E} S^{-T} B S^{-1} v',$$

which results from a Galerkin approximation in the $\mathcal{L}^2$-normalized basis set

$$\mathcal{B}^O_{m_1, m_2} := \{\Phi_1, \ldots, \Phi_{m_1}, \Phi'_{m_1+1}, \ldots \Phi'_M\},$$

where $\Phi'_{\mu_2} = \sum_{\nu=1}^M (S)_{\mu_2, \nu} \Phi_\nu$ for $m_1 + 1 \leq \mu_2 \leq M$. Note particularly that the basis set $\mathcal{B}^O_{m_1, m_2}$ is also partially orthogonal, i.e. the relations $\langle \Phi_{\mu_1}, \Phi'_{\mu_2} \rangle = 0$ hold for $1 \leq \mu_1 \leq m_1$, $m_1 + 1 \leq \mu_2 \leq M$ and $\langle \Phi'_{\mu_2}, \Phi_{\mu_1} \rangle = 0$ for $1 \leq \mu_1 \leq m_1$, $m_1 + 1 \leq \mu_2 \leq M$; compare the right hand side of (4.67).

In this thesis, according to (4.61), we focus on the application of $\mathcal{L}^2$-normalized basis sets $\mathcal{B} = \{\Phi_1, \ldots, \Phi_M\}$ with

$$\Phi_1 = \mathfrak{A}^{(N, M_S)} \left(\vec{\mathbf{y}} \mapsto \prod_{p \in \mathcal{N}} \psi_p(\mathbf{y}_p) \right)$$

in the framework of Algorithm 4.1. Hence, we employ a special case of the transformation matrix $S_{\mathcal{B}}$, where we set the transformation matrix $S_{\mathcal{B}}$ equal to (4.68) with $m_1 = 1$, $m_2 = M - m_1$. Then, in particular, the computation of B_{11}^{-1} is trivial. Now, let $W^{(N,M_S)}$ denote the subspace of span($\mathcal{B}$) with the orthogonal direct sum decomposition span($\mathcal{B}$) = $\mathcal{W}_\emptyset^{(N,M_S)} \oplus W^{(N,M_S)}$.[15] Then, for $\Psi \in \mathcal{B}$ with $\Psi = \sum_{\mu=1}^M v_\mu \Phi_\mu$ we have $\Psi = v_1 \Phi_1 + \sum_{\mu=2}^M v'_\mu \Phi'_\mu$ with $v_1 \Phi_1 \in \mathcal{W}_\emptyset^{(N,M_S)}$ and $\sum_{\mu=2}^M v'_\mu \Phi'_\mu$ in the orthogonal detail space $W^{(N,M_S)}$.[16]

Note that we perform the assembly of the system matrices A and B in parallel in a straightforward way. Here, we distribute the matrix columns over the processors such that each column is stored on one processor only. However, since we deal with full matrices, the information associated with the basis set is replicated on all processors. The data structures which we implemented to deal with the basis sets constructed by the adaptive Algorithm 4.1 are based on the commonly used hash techniques [108, 148].

4.5 Summary

In this chapter we introduced the numerical methods involved in the computation of the minimal energy (with respect to the total spin projection) associated with a molecular Hamiltonian.

First, we discussed the application of the Löwdin rules to deal with partial antisymmetry in an implementation of the Galerkin discretization approach of the electronic Schrödinger equation.

Then, we introduced Gaussian multiscale frames which can be seen as an approximation of Meyer wavelets. Furthermore, we presented a specific type of partially antisymmetric tensor product multiscale many-particle spaces with finite-order weights. Note that the well-known configuration-interaction spaces in quantum chemistry can be particularly identified as a special case of our type of subspaces.

Moreover, we described our h-adaptive scheme to construct a sequence of subspaces by refinement and expansion in real space. Here, we particularly discussed the application of rank-1 approximations in terms of Gaussians and modulated Gaussians for the particle-wise subspace splitting. Furthermore, we presented initial subspaces based on the multiscale Gaussian frame form (4.29) to perform the introduced refinement and expansion scheme. Let us remark that, by using only Gaussians and modulated Gaussian, all the involved one and two particle operator integrals can be given in the form of analytic formulae.

Note that one can apply alternative sets Ng and Ng$^+$ associated with refinement and expansion. For example, a full grid refinement and expansion in the one-particle spaces may be performed by the application of the set $\mathcal{Z}' = \{-1, 0, 1\}^3$ in (4.54) and (4.64), respectively. Moreover, one could construct sets associated with refinement and expansion according to the product structure of two scaled regular hyperbolic crosses (3.33) as considered in Theorem 3.3.

Alternatively, the molecular Hamiltonian H may be regularized by an approximation of the Coulomb interaction potential in (3.1) in terms of Gaussians. This allows for

---

[15]Note that span(Φ_1) = $\tilde{\mathcal{W}}_{\emptyset,\{\Phi_1\}}^{(N,M_S)} = \mathcal{W}_\emptyset^{(N,M_S)}$ according to (4.43) and (4.33).
[16]Here, the squared $\mathcal{L}^2$-norm can be written as $\|\Psi\|^2 = v_1^2 + \|\sum_{\mu=2}^M v'_\mu \Phi'_\mu\|^2$.

4 Numerical Methods

an efficient computation of the matrix entries for various multiscale frames. Then, however, the problem is to perform regularization in such a way, that the error related to the approximation of the Coulomb potential can be controlled and balanced with the approximation error related to the application of a finite subset of the multiscale frame.

5 Numerical Experiments

In this chapter we describe aspects and results of the numerical experiments which we performed with the help of our new adaptive approach according to Algorithm 4.1 introduced in Chapter 4. We discuss the application of our new method to compute approximate energies associated with low states of small molecular systems. Here, we aim at the determination of the total energies of the considered molecular systems in the so-called chemical accuracy. Commonly, an accuracy of approximately 1 kcal/mol ≈ 1.595_{-3} hartree is regarded as chemical accuracy [187]. Note that the main task here is to efficiently describe electron correlation [175]. To demonstrate that our method allows for an efficient representation of basic electron-nuclei and electron-electron cusps, we first consider the test cases of the ground-state as well as the lowest fully antisymmetric state of the He atom and the H_2 molecule. Additionally, we apply our new approach to several small atomic and diatomic systems with up to six electrons involved.

In particular, we compare our numerical results with those obtained by a Full-CI approach (4.46) where *linear combination of atomic orbitals* (LCAO) basis sets [47, 189] are employed, which are particularly common in quantum chemistry [170]. We further compare our numerical results to that obtained by other widely used methods in quantum chemistry like the variational quantum Monte Carlo method and diffusion quantum Monte Carlo method. Here, we only consider the case of a single-determinant Jastrow-Slater function as trial wave function; see e.g. [21, 127, 177]. Moreover, we compare our results to the coupled cluster method, i.e. the CCSD(T) method as implemented in the quantum chemistry software package PSI3 [36]; see e.g. [86, 161, 175].

In the following we note some specific aspects of the application of our adaptive scheme according to Algorithm 4.1 in the framework of this thesis.

We employ one-particle functions given by non-orthogonal unrestricted Hartree-Fock orbitals as in (4.36) for the underlying particle-wise subspace splitting, i.e. we set $\{\psi_p = \tilde{\psi}_p^{HF}\}_{p \in \mathcal{N}}$. We represent each $\tilde{\psi}_p^{HF}$ in terms of linear combinations of s-type and p-type atomic orbitals of the form (4.37) and (4.38) centered at a nucleus. We only employ sums of nine one-particle (modulated) Gaussians, i.e. we set $Q = 9$ in (4.37) and (4.38). Here, to shorten notation, s- and p-type orbitals which are centered at the origin are denoted by $o^{(s)}$ and $o^{(p)}$, respectively. For each studied molecular system, we give the specific choice of linear combinations of orbitals.

An initial system for Algorithm 4.1 is constructed by using a basis set $\mathfrak{b}_{\sigma,c}^{(L)}$ as in (4.51). Note here that $\mathfrak{b}_{\sigma,c}^{(L)}$ is a subset of the second order Gaussian multiscale frame $\mathfrak{b}_{\sigma,c}$ according to (4.29). Furthermore, for each studied molecular system we set $\sigma = 1$.

We set the parameter δ to a value of $1/100$ kcal/mol ≈ 1.59_{-5} hartree in the coarsening step of $\mathcal{B}_{three}^{[0]} \cup \mathcal{B}_{other}^{[0]}$ to obtain an initial system $\mathcal{B}^{[0]}$ for Algorithm 4.1; compare also (4.61), (4.62) and (4.63). Moreover, we apply a threshold parameter equal to $\varepsilon_\kappa = 2^{-\kappa}$ in Algorithm 4.1.

In the following we give some notational aspects. In particular, we give all quantities in atomic units, e.g. energy values in hartree and distance values in bohr. We denote the approximation error $|E_\kappa^{(1)} - \tilde{E}_\kappa^{(1)}|$ associated with step 5 in Algorithm 4.1 by e_κ. Moreover, the respective partial energy terms associated with the sum of subspaces given in (4.48) are denoted by E_κ^{zero}, E_κ^{one}, $E_\kappa^{two\uparrow\downarrow}$, $E_\kappa^{two\uparrow\uparrow}$ and E_κ^{three}. The corresponding numbers of degrees of freedom are denoted by M_κ^{zero}, M_κ^{one}, $M_\kappa^{two\uparrow\downarrow}$, $M_\kappa^{two\uparrow\uparrow}$ and M_κ^{three}. Additionally, in the case of $N > 3$ we denote the energy term and the number of degrees of freedom associated with the set $\mathcal{B}_{other}^{[0]}$ as in (4.62) by E_κ^{other} and M_κ^{other}. Furthermore, we denote the piecewise gradients in the log-log variant with respect to the number of degrees of freedom by $s_\kappa^M = \frac{\log(e_\kappa) - \log(e_{\kappa-1})}{\log(M_\kappa) - \log(M_{\kappa-1})}$. Analogously, the piecewise gradients in the log-log variant with respect to the threshold parameter are denoted by $s_\kappa^\varepsilon = \frac{\log(e_\kappa) - \log(e_{\kappa-1})}{\log(\varepsilon_\kappa) - \log(\varepsilon_{\kappa-1})}$.

Furthermore, according to (3.5), we denote the total energy by E_{tot}, i.e. the sum of the energy associated with the electronic Hamiltonian and the energy term associated with nuclei-nuclei interaction V_{nn}. Correspondingly, the sum of the computed approximate energy term $\tilde{E}_{\kappa_{max}}^{(1)}$ and V_{nn} is denoted by $\tilde{E}_{tot}$. Also, we denote the total energy associated with the Hartree-Fock limit by E_{tot}^{HF} and the total energy associated with the rank-1 approximation according to (4.36) by $\tilde{E}_{tot}^{HF}$. The *exact* energy values E_{tot} and the exact Hartree-Fock limits E_{tot}^{HF} are taken from literature.

Note that the so-called correlation energy is defined by $E_{tot}^{HF} - E_{tot}$; see e.g. [170]. We denote the proportion of the correlation energy with respect to the approximate energy $\tilde{E}_{tot}$ by

$$E_c := \frac{\tilde{E}_{tot} - E_{tot}^{HF}}{E_{tot} - E_{tot}^{HF}}, \qquad (5.1)$$

where we usually give E_c in percentage.

Note finally that all numerical experiments are performed on the cluster computer Himalaya.[1] This parallel computing system consists of 256 *Intel Xeon EM 64 T 3.2 GHz* processors interconnected by a *Myrinet XP*.

5.1 He atom, H_2 and He_2^+ molecules

First, we consider the ground-state energy of the He atom with $M_S = 0$. For the rank-1 approximation in (4.36) we apply an $\mathcal{L}^2$-normalized partially antisymmetric function of the form $\tilde{\Psi}_{(2,0)}^{HF} = C\mathfrak{A}^{(2,0)}(\psi_{1(\uparrow)}^{He} \otimes \psi_{2(\downarrow)}^{He})$, where C denotes a normalization constant. Here, we employ an s-type orbital for the spin-up one-particle function $\psi_{1(\uparrow)}^{He}$ and set the spin-down one-particle function $\psi_{2(\downarrow)}^{He}$ equal to $\psi_{1(\uparrow)}^{He}$; see also Figure 5.1(a). Moreover, with regard to the initial set $\mathfrak{b}_{1,c}^{(L)}$, we use a value of $\frac{1}{2}$ for the parameter c and a value of 7 for the parameter L. A visualization of the computed approximations to the wave function of the He atom is given in Figure 5.3 and Figure 5.4. In the case of the He atom execution times associated with the eigenvalue solver routines within application of Algorithm 4.1 are given in Figure 5.2. Our results suggest that the complexities of the parallel eigenvalue solvers discussed in Section 4.1.2, i.e. LOBPCG, Krylov-Schur and PARPACK, are all of the same order for this system. Note that a log-log fit results in complexities with respect to the number of particles of order 2, which in particular can be expected due to the full system matrices associated with the general Hermitian

---
[1] See http://wissrech.ins.uni-bonn.de/research/himalaya/hardware.html for more details on the cluster computer Himalaya.

5.1 He atom, H_2 and He_2^+ molecules

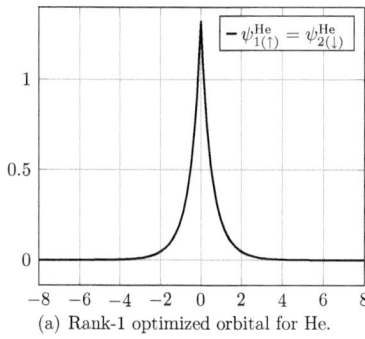
(a) Rank-1 optimized orbital for He.

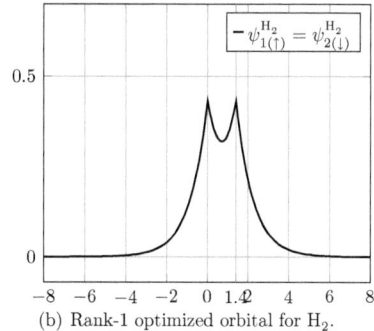
(b) Rank-1 optimized orbital for H_2.

Figure 5.1: Visualization of the one-particle functions corresponding to the rank-1 approximation in (4.36) for the ground-state of He and H_2 along the x_1 axis.

eigenvalue problem (4.1). However, in the case of the implementation of the LOBPCG method the involved constant seems to be the smallest. In the following, we therefore invoke the LOBPCG method as implemented in the BLOPEX package together with a Cholesky decomposition as implemented in the software package PLAPACK; compare Section 4.1.2.

Additionally, we approximately compute the ground-state energies of the H_2 and the He_2^+ molecules. The technical details are given in the following: We consider the H_2 molecule with $M_S = 0$ and with bond distance $R_{HH} = 1.4$ bohr, i.e. $R_1 = 0$ and $R_2 = (R_{HH}, 0, 0)^T$. For the rank-1 approximation in (4.36) we apply an $\mathcal{L}^2$-normalized partially antisymmetric function of the form $\tilde{\Psi}_{(2,0)}^{HF} = C\mathfrak{A}^{(2,0)}(\psi_{1(\uparrow)}^{H_2} \otimes \psi_{2(\downarrow)}^{H_2})$. For the spin-up one-particle function, we employ $\psi_{1(\uparrow)}^{H_2}(\mathbf{x}) = o^{(s)}(\mathbf{x}) + o^{(s)}(\mathbf{x} - \mathbf{R}_2)$, where $o^{(s)}$ denotes an s-type orbital centered at the origin $\mathbf{R}_1 = 0$. We set the spin-down one-particle function $\psi_{2(\downarrow)}^{H_2}$ equal to $\psi_{1(\uparrow)}^{H_2}$; see also Figure 5.1(b). Furthermore, we employ a value of $\frac{2}{R_{HH}}$ for the parameter c and a value of 5 for the parameter L. In the case of the He_2^+ molecule with $M_S = \pm 1/2$ we consider a bond distance of $R_{HeHe^+} = 2.042$ bohr, where without loss of generality we only focus on the case of $M_S = 1/2$ in the following. For the rank-1 approximation in (4.36) we apply an $\mathcal{L}^2$-normalized partially antisymmetric function of the form $\tilde{\Psi}_{(3,1/2)}^{HF} = C\mathfrak{A}^{(3,1/2)}(\psi_{1(\uparrow)}^{He_2^+} \otimes \psi_{2(\uparrow)}^{He_2^+} \otimes \psi_{3(\downarrow)}^{He_2^+})$. For the spin-up one-particle functions we employ $\psi_{1(\uparrow)}^{He_2^+}(\mathbf{x}) = o_1^{(s)}(\mathbf{x})$ and $\psi_{2(\uparrow)}^{He_2^+}(\mathbf{x}) = o_1^{(s)}(\mathbf{x} - \mathbf{R}_2)$, where we set $\psi_{3(\downarrow)}^{He_2^+}(\mathbf{x}) = o_2^{(s)}(\mathbf{x}) + o_2^{(s)}(\mathbf{x} - \mathbf{R}_2)$ for the spin-down one-particle function. Here, $o_1^{(s)}$ and $o_2^{(s)}$ denote s-type orbitals centered at the origin. Moreover, we apply a value of $\frac{2}{R_{HeHe^+}}$ for the parameter c and a value of 6 for the parameter L.

Our numerical results according to He, H_2 and He_2^+ are given in Table 5.1, Table 5.2 and Table 5.3, respectively. In addition, the results are summarized in Figure 5.5 and compared with results obtained by other methods in Table 5.4.

We see that our method is indeed convergent if the number of degrees of freedom M

5 Numerical Experiments

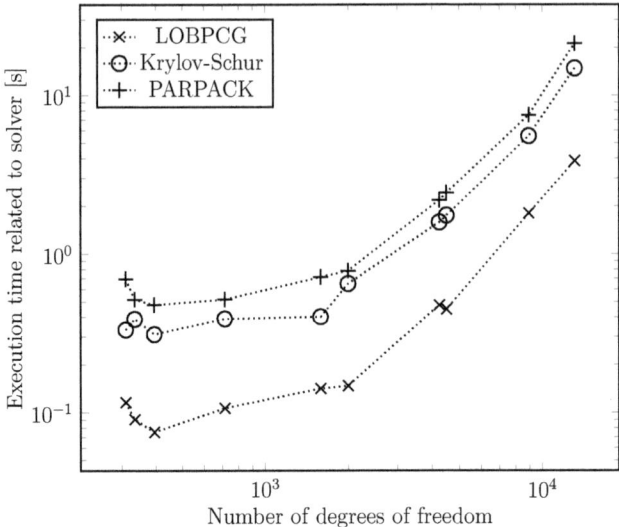

Figure 5.2: Execution times in seconds associated with solver routines versus the number of degrees of freedom M associated with respective subspaces created by Algorithm 4.1 in the case of the computation of an approximation of the He ground-state energy. Here, we present the results for 32 processors on the Himalaya cluster computer for different parallel eigenvalue solvers as discussed in Section 4.1.2. Note that for $\kappa \leq 1$, we use the computed ground-state in span$(\mathcal{B}^{[\kappa-1]})$ as the initial state for the iteration according to the invoked eigenvalue solver for the solution in span$(\mathcal{B}^{[\kappa]})$.

increases with the number of steps κ of Algorithm 4.1. In particular, for all systems chemical accuracy ($\approx 1\,\mathrm{kcal/mol}$) is reached and a proportion of the correlation energies of more than 97 % is obtained, e.g. 99 % for He. Moreover, the measured orders of convergence are predominantly better than the order of $-1/3$ for rising M. Note that, due to Theorem 3.2 and relation (3.20), one can expect an order of $-1/3$ in the case of a linear approximation scheme. Furthermore, our results are in the same range as the results obtained by CCSD(T) and Full-CI computations using the PSI3 package [36] with the cc-pV5Z basis set [189]. Moreover, let us note that in the case of the three-electron system He_2^+ the energy terms E_κ^{zero}, E_κ^{one} and $E_\kappa^{two\uparrow\downarrow}$ are of the largest size in the particle-wise decomposition.

5.1 He atom, H_2 and He_2^+ molecules

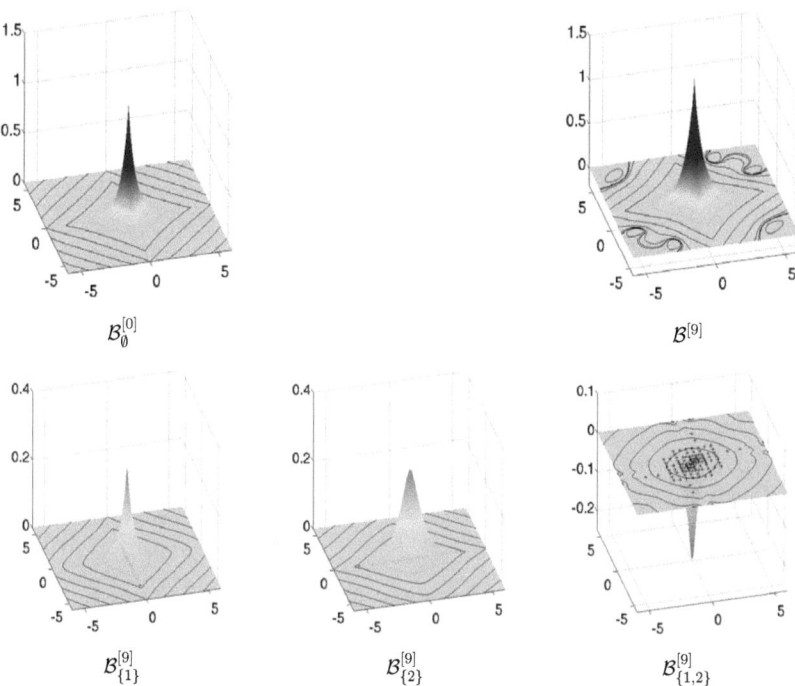

Figure 5.3: Visualization concerning the subspace splitting (4.48) of the approximately computed ground state wave function of the He atom according to Algorithm 4.1. The isolines are in logarithmic scale and the localization peaks of basis functions are marked by a plus. Note that we have the inclusions $\text{span}\{\mathcal{B}_{\emptyset}^{[0]}\} \subset \mathcal{V}^{zero}$, $\text{span}\{\mathcal{B}_{\{1\}}^{[9]} \cup \mathcal{B}_{\{2\}}^{[9]}\} \subset \mathcal{V}^{one}$, $\text{span}\{\mathcal{B}_{\{1,2\}}^{[9]}\} \subset \mathcal{V}^{two\uparrow\downarrow}$ and $\text{span}\{\mathcal{B}^{[9]}\} \subset \mathcal{V}_{three}^{(2,0)}$.

5 Numerical Experiments

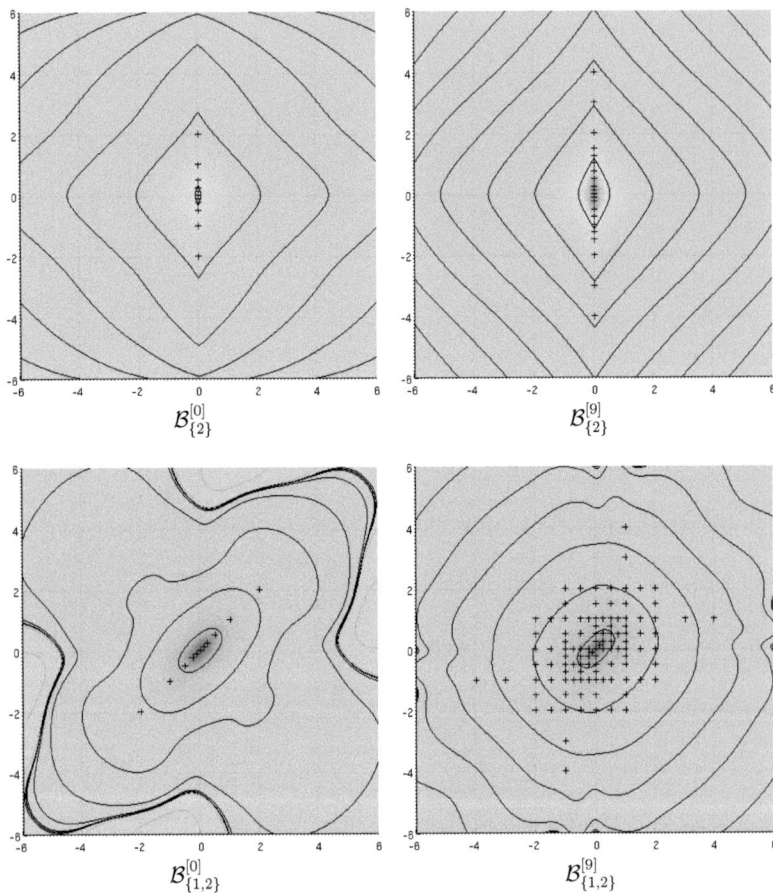

Figure 5.4: Visualization of a projection of the computed approximations of the wave function of the He atom according to Algorithm 4.1, where the initial sets on the left are given according to (4.52), (4.53) and (4.63). The isolines are in logarithmic scale and the localization peaks of basis functions are marked by a plus.

Table 5.1: Numerical results of Algorithm 4.1 for the He atom.

κ	M_κ	$\tilde{E}_\kappa^{(1)}$	e_κ	s_κ^M	s_κ^ε
1	310	-2.897506	6.21_{-3}		
4	334	-2.899675	4.04_{-3}		
5	394	-2.899676	4.04_{-3}		
6	712	-2.901834	1.89_{-3}	-1.29	1.10
7	1,588	-2.902622	1.10_{-3}	-0.67	0.78
8	1,996	-2.902752	9.68_{-4}	-0.55	0.18
9	4,500	-2.902877	8.43_{-4}	-0.17	0.20
10	13,104	-2.903085	6.35_{-4}	-0.26	0.41
11	32,608	-2.903298	4.22_{-4}	-0.45	0.59

κ	E_κ^{zero}	E_κ^{one}	$E_\kappa^{two\uparrow\downarrow}$	M_κ^{one}	$M_\kappa^{two\uparrow\downarrow}$
1	-2.56	-0.63	0.30	198	111
4	-2.29	-1.17	0.56	198	135
5	-2.30	-1.16	0.56	258	135
6	-1.86	-2.05	1.00	258	453
7	-1.88	-2.02	1.00	390	1,197
8	-1.91	-1.97	0.98	486	1,509
9	-1.82	-2.13	1.05	966	3,533
10	-1.73	-2.33	1.16	1,290	11,813
11	-1.64	-2.49	1.23	2,730	29,877

Table 5.2: Numerical results of Algorithm 4.1 for the H_2 molecule.

κ	M_κ	$\tilde{E}_\kappa^{(1)}$	e_κ	s_κ^M	s_κ^ε
1	368	-1.880568	8.19_{-3}		
4	416	-1.883199	5.56_{-3}		
5	644	-1.884173	4.58_{-3}	-0.44	0.28
6	1,348	-1.885517	3.24_{-3}	-0.47	0.50
7	8,742	-1.887183	1.57_{-3}	-0.39	1.04
8	19,808	-1.888046	7.09_{-4}	-0.97	1.15

κ	E_κ^{zero}	E_κ^{one}	$E_\kappa^{two\uparrow\downarrow}$	M_κ^{one}	$M_\kappa^{two\uparrow\downarrow}$
1	-1.44	-0.79	0.35	306	61
4	-1.33	-1.02	0.46	306	109
5	-1.27	-1.09	0.48	354	289
6	-1.16	-1.32	0.60	482	865
7	-1.01	-1.60	0.71	1,114	7,627
8	-0.90	-1.83	0.84	1,594	18,213

5 Numerical Experiments

Table 5.3: Numerical results of Algorithm 4.1 for the He_2^+ molecular ion.

κ	M_κ	$\tilde{E}_\kappa^{(1)}$	e_κ	s_κ^M	s_κ^ε
1	1,412	−6.935416	1.81_{-2}		
3	1,460	−6.939936	1.36_{-2}		
4	1,472	−6.942464	1.10_{-2}	−25.15	0.30
5	1,980	−6.946905	6.60_{-3}	−1.74	0.74
6	4,036	−6.949023	4.48_{-3}	−0.54	0.56
7	7,522	−6.950432	3.07_{-3}	−0.61	0.54
8	16,757	−6.951912	1.59_{-3}	−0.82	0.95

κ	E_κ^{zero}	E_κ^{one}	$E_\kappa^{two\uparrow\downarrow}$	$E_\kappa^{two\uparrow\uparrow}$	E_κ^{three}	M_κ^{one}	$M_\kappa^{two\uparrow\downarrow}$	$M_\kappa^{two\uparrow\uparrow}$	M_κ^{three}
1	−6.83	0.18	−0.29	-4.06_{-2}	$ 4.60_{-2}$	455	444	330	182
3	−7.06	0.66	−0.55	-7.13_{-3}	$ 1.53_{-2}$	455	492	330	182
4	−7.48	1.22	−0.70	$ 4.90_{-3}$	$ 1.27_{-2}$	467	492	330	182
5	−6.18	−1.48	0.70	$ 2.31_{-2}$	-5.36_{-3}	495	972	330	182
6	−5.52	−2.61	1.16	$ 3.01_{-2}$	-1.14_{-2}	823	2,700	330	182
7	−6.47	−0.66	0.17	$ 3.32_{-2}$	-1.25_{-2}	1,243	5,766	330	182
8	−6.90	0.04	−0.18	$ 9.38_{-2}$	-1.36_{-2}	2,553	12,618	1,403	182

Table 5.4: Comparison of numerical results for the He atom, the H_2 molecule and the He_2^+ molecular ion. The Hartree-Fock energy $E_{tot}^{HF}(\text{He}_2^+)$ is obtained by a restricted open-shell Hartree-Fock (ROHF) computation using the PSI3 package [36] with the cc-pV5Z basis set [47, 189]. The energies $E_{tot}^{CCSD(T)}$ and E_{tot}^{FCI} are obtained by a CCSD(T) and a Full-CI computation, respectively, using the PSI3 package [36] with the cc-pV5Z basis set [189].

	He		H_2		He_2^+	
$\tilde{E}_{tot}$	−2.90330		−1.17376		−4.99305	
E_{tot}	−2.90372	[38]	−1.17447	[111]	−4.99464	[28]
$\tilde{E}_{tot}^{HF}$	−2.86166		−1.12854		−4.91647	
E_{tot}^{HF}	−2.86168	[38]	−1.13366	[172]	−4.92285	
$E_c[\%]$	99.00		98.26		97.78	
$E_{tot}^{CCSD(T)}$	−2.90315		−1.17422		−4.99268	cc-pV5Z
E_{tot}^{FCI}	−2.90315		−1.17422		−4.99273	cc-pV5Z

5.1 He atom, H_2 and He_2^+ molecules

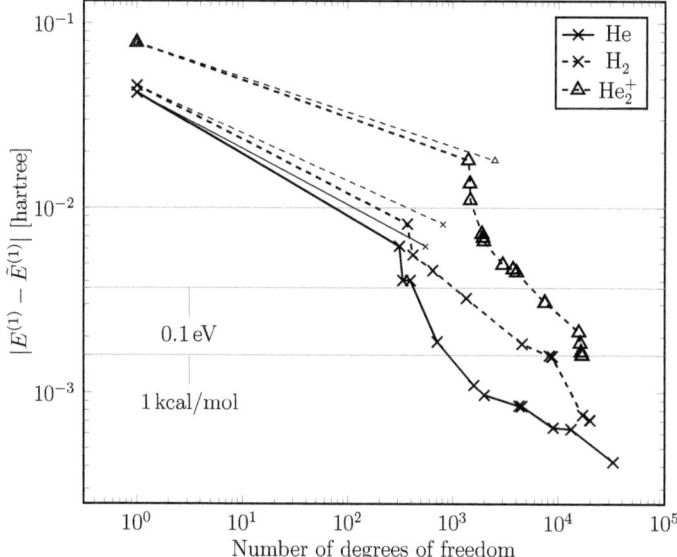

(a) Error versus number of degrees of freedom, where each step 3 in Algorithm 4.1 is taken into account; compare (4.65). The small marker denotes the initial system $\mathcal{B}^{[0]}_{three} \cup \mathcal{B}^{[0]}_{other}$ which is used in the coarsening step to obtain $\mathcal{B}^{[0]}$; compare (4.63).

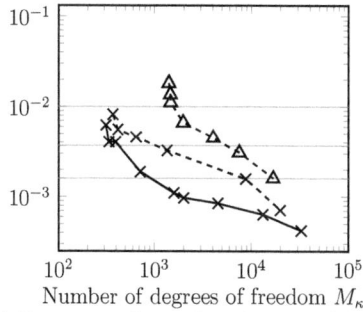

(b) Error versus the number of degrees of freedom, where each $|E^{(1)}_\kappa - \tilde{E}^{(1)}_\kappa|$ and M_κ associated with step 5 in Algorithm 4.1 is taken into account.

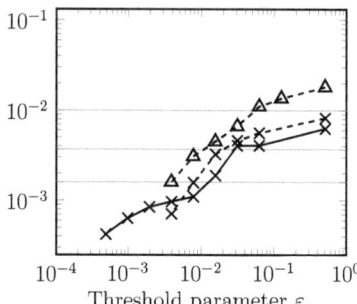

(c) Error versus the threshold parameter, where each $|E^{(1)}_\kappa - \tilde{E}^{(1)}_\kappa|$ and ε_κ associated with step 5 in Algorithm 4.1 is taken into account.

Figure 5.5: Error of the approximately computed energy of the ground-state of the He atom, of the H_2 molecule and of the He_2^+ molecular ion.

5.2 Lowest fully antisymmetric states of He, H₂ and Li

Let us now perform our new method to approximately compute the energy associated with the lowest fully antisymmetric state of the He atom, the H$_2$ molecule and the Li atom, respectively.

We give the technical details in the following: We consider the lowest triplet state of the He atom, i.e. $N = 2$ and $M_S = \pm 1$, which is also denoted by (^3S)He. Without loss of generality we focus on the case of $M_S = 1$ only. Then, for the rank-1 approximation in (4.36) we apply an $\mathcal{L}^2$-normalized fully antisymmetric function of the form $\tilde{\Psi}_{(2,1)}^{HF} = C\mathfrak{A}^{(2,1)}(\psi_{1(\uparrow)}^{(^3S)\text{He}} \otimes \psi_{2(\uparrow)}^{(^3S)\text{He}})$. Here, for each of the spin-up one-particle functions $\psi_{1(\uparrow)}^{(^3S)\text{He}}$ and $\psi_{2(\uparrow)}^{(^3S)\text{He}}$ we employ an s-type orbital centered at the origin. Moreover, we use a value of $\frac{1}{4}$ for the parameter c and a value of 7 for the parameter L. We consider the lowest triplet state of the H$_2$ molecule in the case of a distance $R_{3\text{HH}} = 2.0\,\text{bohr}$, i.e. $N = 2$, $M_S = \pm 1$, $\mathbf{R}_1 = \mathbf{0}$ and $\mathbf{R}_2 = (R_{3\text{HH}}, 0, 0)^T$. Note that the lowest triplet state of the H$_2$ is also denoted by $(b^3\Sigma_u^+)$H$_2$. Without loss of generality we again focus on the case $M_S = 1$ only. Then, for the rank-1 approximation in (4.36) we apply an $\mathcal{L}^2$-normalized fully antisymmetric function of the form $\tilde{\Psi}_{(2,1)}^{HF} = C\mathfrak{A}^{(2,1)}(\psi_{1(\uparrow)}^{(b^3\Sigma_u^+)\text{H}_2} \otimes \psi_{2(\uparrow)}^{(b^3\Sigma_u^+)\text{H}_2})$. For the spin-up one-particle functions we set $\psi_{1(\uparrow)}^{(b^3\Sigma_u^+)\text{H}_2}(\mathbf{x}) = o_1^{(s)}(\mathbf{x}) + o_1^{(s)}(\mathbf{x} - \mathbf{R}_2)$ and $\psi_{2(\uparrow)}^{(b^3\Sigma_u^+)\text{H}_2} = o_2^{(s)}(\mathbf{x}) - o_2^{(s)}(\mathbf{x} - \mathbf{R}_2)$. Here, $o_1^{(s)}$ and $o_2^{(s)}$ denote s-type orbital centered at $\mathbf{R}_1 = \mathbf{0}$, respectively. Moreover, we employ a value of $\frac{2}{R_{3\text{HH}}}$ for the parameter c and a value of 5 for the parameter L. Finally, in the case of the lowest quartet state of the Li

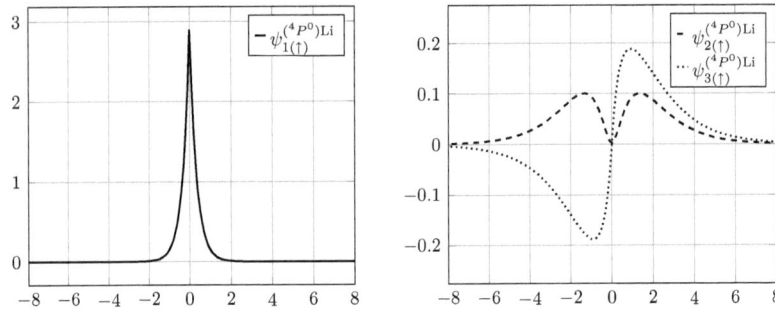

Figure 5.6: Visualization of the one-particle functions corresponding to the rank-1 approximation in (4.36) for the lowest quartet state of Li, i.e. $^4P^0$. Here, we plot the one-particle functions along the x_1 axis. Note that all one-particle functions are rotationally symmetric s-type orbitals except for $\psi_{3(\uparrow)}^{(^4P^0)\text{Li}}$, which is a p_1-type orbital; compare also (4.37) and (4.38).

atom, i.e. $N = 3$ and $M_S = \pm 3/2$, which is also denoted by $(^4P^0)$Li, we focus on the case of $M_S = 3/2$ only. For the rank-1 approximation in (4.36) we apply an $\mathcal{L}^2$-normalized fully antisymmetric function of the form $\tilde{\Psi}_{(3,3/2)}^{HF} = C\mathfrak{A}^{(3,3/2)}(\psi_{1(\uparrow)}^{(^4P^0)\text{Li}} \otimes \psi_{2(\uparrow)}^{(^4P^0)\text{Li}} \otimes \psi_{3(\uparrow)}^{(^4P^0)\text{Li}})$.

5.2 Lowest fully antisymmetric states of He, H$_2$ and Li

Here, for each of the one-particle functions $\psi_{1(\uparrow)}^{(^4P^0)\text{Li}}$ and $\psi_{2(\uparrow)}^{(^4P^0)\text{Li}}$ we employ an s-type orbital centered at the origin. For the one-particle function $\psi_{3(\uparrow)}^{(^4P^0)\text{Li}}$ we use an p-type orbital centered at the origin; compare Figure 5.6. Moreover, we apply a value of $\frac{1}{4}$ for the parameter c and a value of 7 for the parameter L.

The corresponding numerical results are given in Table 5.5, Table 5.6 and Table 5.7. Moreover, we compare our results with those computed by other methods in Table 5.8 and we summarize them in Figure 5.7.

We see that there is convergence and that for all systems chemical accuracy is reached. In particular, an error in the range of ten micro hartree ($\approx 10^{-5}$ hartree) is obtained for the two-electron systems (^3S)He and ($b^3\Sigma_u^+$)H$_2$. Moreover, for the two-electron systems proportions of the correlation energies of more than 99 % are obtained, whereas for the three-electron system a value of more than 96 % is still reached. Furthermore, in most cases the measured orders of convergence are better than an order of $-2/3$ for rising M. Here, since we consider fully antisymmetric states, one can expect an order of $-2/3$ in the case of a linear approximation scheme due to Theorem 3.2 and relation (3.20). Furthermore, the approximation errors of our results are smaller than those obtained by CCSD(T) and Full-CI computations using the PSI3 package [36] with the cc-pV5Z basis set [189]. Note further that in the case of the two-electron system (^3S)He and ($b^3\Sigma_u^+$)H$_2$ the partial energy terms E_κ^{zero} are of the largest size within the particle-wise decomposition.

Table 5.5: Numerical results of Algorithm 4.1 for the lowest triplet state of the He atom (i.e. (^3S)He, $N = 2$, $M_S = \pm 1$).

κ	M_κ	$\tilde{E}_\kappa^{(1)}$	e_κ	s_κ^M	s_κ^ε
1	141	-2.174698	5.31_{-4}		
5	147	-2.174698	5.31_{-4}		
7	741	-2.174883	3.46_{-4}		
8	1,195	-2.174973	2.56_{-4}	-0.62	0.43
9	4,665	-2.175189	3.99_{-5}	-1.37	2.68
10	7,087	-2.175206	2.26_{-5}	-1.36	0.82
11	11,481	-2.175217	1.24_{-5}	-1.23	0.86
12	20,386	-2.175220	9.22_{-6}	-0.52	0.43

κ	E_κ^{zero}	E_κ^{one}	$E_\kappa^{two\uparrow\uparrow}$	M_κ^{one}	$M_\kappa^{two\uparrow\uparrow}$
1	-2.21	0.02	2.07_{-2}	76	64
5	-2.23	0.03	2.08_{-2}	82	64
7	-2.16	-0.11	9.60_{-2}	220	520
8	-2.19	-0.14	1.57_{-1}	336	858
9	-2.40	0.09	1.35_{-1}	751	3,913
10	-2.37	0.07	1.28_{-1}	1,401	5,685
11	-2.21	-0.17	1.98_{-1}	2,266	9,214
12	-2.31	-0.10	2.26_{-1}	3,753	16,632

Table 5.6: Numerical results of Algorithm 4.1 for the lowest triplet state of the H_2 molecule (i.e. $(b^3\Sigma_u^+)H_2$, $N = 2$, $M_S = \pm 1$).

κ	M_κ	$\tilde{E}_\kappa^{(1)}$	e_κ	s_κ^M	s_κ^ε
1	350	-1.393607	3.47_{-3}		
6	1,069	-1.396413	6.63_{-4}		
7	2,170	-1.396673	4.04_{-4}	-0.70	0.72
8	6,152	-1.396920	1.56_{-4}	-0.91	1.37
9	15,803	-1.397005	7.15_{-5}	-0.83	1.13
10	32,147	-1.397040	3.59_{-5}	-0.97	0.99

κ	E_κ^{zero}	E_κ^{one}	$E_\kappa^{two\uparrow\uparrow}$	M_κ^{one}	$M_\kappa^{two\uparrow\uparrow}$
1	-1.36	-0.05	1.75_{-2}	251	98
6	-1.28	-0.19	7.07_{-2}	333	735
7	-1.35	-0.07	1.71_{-2}	632	1,537
8	-1.42	0.03	-2.89_{-3}	1,469	4,682
9	-1.48	0.12	-3.78_{-2}	2,487	13,315
10	-1.48	0.11	-3.51_{-2}	4,302	27,844

Table 5.7: Numerical results of Algorithm 4.1 for the lowest quartet state of the Li atom (i.e. $(^4P^0)Li$, $N = 3$, $M_S = \pm 3/2$).

κ	M_κ	$\tilde{E}_\kappa^{(1)}$	e_κ	s_κ^M	s_κ^ε
1	901	-5.363903	4.11_{-3}		
4	942	-5.364110	3.90_{-3}		
5	1,227	-5.365122	2.89_{-3}	-1.14	0.43
6	1,725	-5.365974	2.04_{-3}	-1.03	0.50
7	3,631	-5.366554	1.46_{-3}	-0.45	0.48
8	6,009	-5.366696	1.31_{-3}	-0.20	0.15
9	13,306	-5.367261	7.49_{-4}	-0.71	0.81
10	23,706	-5.367640	3.71_{-4}	-1.22	1.01

κ	E_κ^{zero}	E_κ^{one}	$E_\kappa^{two\uparrow\uparrow}$	M_κ^{one}	$M_\kappa^{two\uparrow\uparrow}$
1	-3.67	-2.15	4.58_{-1}	314	586
4	-2.81	-3.08	5.22_{-1}	355	586
5	-2.29	-4.11	1.03_{0}	463	763
6	-2.57	-3.67	8.83_{-1}	667	1,057
7	-2.31	-4.56	1.50_{0}	969	2,661
8	-2.67	-3.42	7.22_{-1}	1,433	4,575
9	-3.81	-2.58	1.02_{0}	2,331	10,974
10	-3.50	-3.44	1.57_{0}	3,255	20,450

5.2 Lowest fully antisymmetric states of He, H_2 and Li

Table 5.8: Comparison of numerical results for the lowest triplet state of the He atom ((^3S)He), the lowest triplet state of the H_2 molecule (($b^3\Sigma_u^+$)H_2) and the lowest quartet state of the Li atom (($^4P^0$)Li). The Hartree-Fock energy $E_{tot}^{HF}(^3H_2)$ is obtained by a restricted open-shell Hartree-Fock (ROHF) computation using the PSI3 [36] package with the cc-pV5Z basis set [47, 189]. The energies $E_{tot}^{CCSD(T)}$ and E_{tot}^{FCI} are obtained by a CCSD(T) and a Full-CI computation, respectively, using the PSI3 package [36] with the cc-pV5Z basis set [189].

	(^3S)He		($b^3\Sigma_u^+$)H_2		($^4P^0$)Li	
$\tilde{E}_{tot}$	−2.17522		−0.89704		−5.36764	
E_{tot}	−2.17523	[140]	−0.89708	[27]	−5.36801	[11]
$\tilde{E}_{tot}^{HF}$	−2.17424		−0.89057		−5.35830	
E_{tot}^{HF}	−2.17420	[101]	−0.89288		−5.35830	[101]
$E_c[\%]$	99.10		99.14		96.18	
$E_{tot}^{CCSD(T)}$	−2.04194		−0.89691		−5.24319	cc-pV5Z
E_{tot}^{FCI}	−2.04194		−0.89691		−5.24320	cc-pV5Z

5 Numerical Experiments

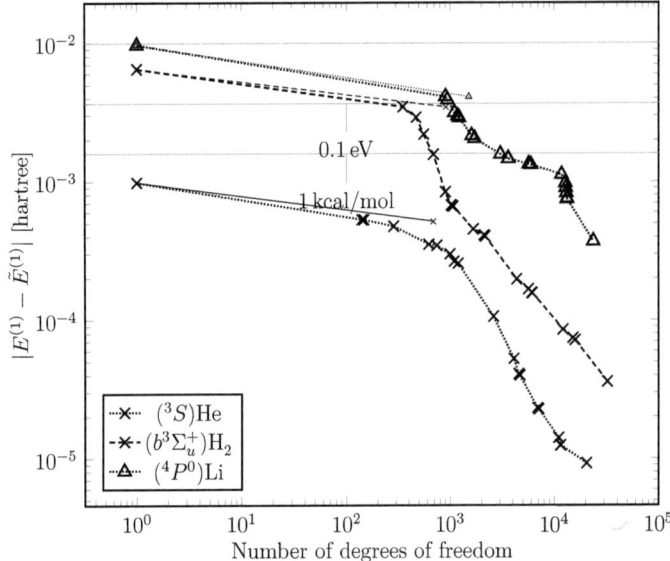

(a) Error versus number of degrees of freedom, where each step 3 in Algorithm 4.1 is taken into account; compare (4.65). The small marker denotes the initial system $\mathcal{B}^{[0]}_{three} \cup \mathcal{B}^{[0]}_{other}$ which is used in the coarsening step to obtain $\mathcal{B}^{[0]}$; compare (4.63).

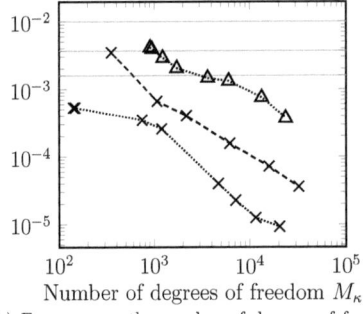

(b) Error versus the number of degrees of freedom, where each $|E^{(1)}_\kappa - \tilde{E}^{(1)}_\kappa|$ and M_κ associated with step 5 in Algorithm 4.1 is taken into account.

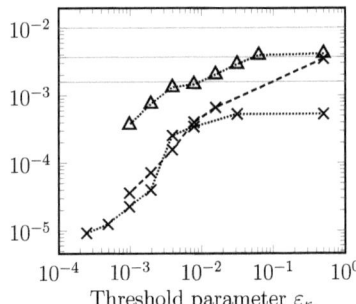

(c) Error versus the threshold parameter, where each $|E^{(1)}_\kappa - \tilde{E}^{(1)}_\kappa|$ and ε_κ associated with step 5 in Algorithm 4.1 is taken into account.

Figure 5.7: Error of the approximately computed energy of the fully antisymmetric states (^3S)He, $(b^3\Sigma^+_u)$H$_2$ and $(^4P^0)$Li.

5.3 Li, Be, B and C atoms

So far we have applied our new method to systems with up to three electrons. Now, we study atomic systems with up to six electrons, i.e. the Li atom, the Be atom, the B atom and the C atom.

We give the technical details in the following: Without loss of generality we focus on the case of $M_S = 1/2$ only for the ground-state of the Li atom. Here, for the rank-1 approximation in (4.36) we apply an $\mathcal{L}^2$-normalized partially antisymmetric function of the form $\tilde{\Psi}^{HF}_{(3,1/2)} = C\mathfrak{A}^{(3,1/2)}(\psi^{Li}_{1(\uparrow)} \otimes \psi^{Li}_{2(\uparrow)} \otimes \psi^{Li}_{3(\downarrow)})$, where for each of the one-particle functions $\psi^{Li}_{1(\uparrow)}$, $\psi^{Li}_{2(\uparrow)}$ and $\psi^{Li}_{3(\downarrow)}$ we employ an s-type orbital centered at the origin. We apply a value of one for the parameter c and a value of 6 for the parameter L. In the case of the Be atom with $M_S = 0$ we apply for the rank-1 approximation in (4.36) an $\mathcal{L}^2$-normalized partially antisymmetric function of the form $\tilde{\Psi}^{HF}_{(4,0)} = C\mathfrak{A}^{(4,0)}(\psi^{Be}_{1(\uparrow)} \otimes \psi^{Be}_{2(\uparrow)} \otimes \psi^{Be}_{3(\downarrow)} \otimes \psi^{Be}_{3(\downarrow)})$. Here, we set $\psi^{Be}_{1(\uparrow)} = \psi^{Be}_{3(\downarrow)} = o_1^{(s)}$ and $\psi^{Be}_{2(\uparrow)} = \psi^{Be}_{4(\downarrow)} = o_2^{(s)}$, where o_1^s and o_2^s denote s-type orbitals. Moreover, we use a value of $\frac{1}{4}$ for the parameter c and a value of 8 for the parameter L. In the case of the ground-state of the B atom we focus on the case of $M_S = 1/2$ only, where for the rank-1 approximation in (4.36) we apply an $\mathcal{L}^2$-normalized partially antisymmetric function of the form $\tilde{\Psi}^{HF}_{(3,1/2)} = C\mathfrak{A}^{(5,1/2)}(\psi^{B}_{1(\uparrow)} \otimes \psi^{B}_{2(\uparrow)} \otimes \psi^{B}_{3(\uparrow)} \otimes \psi^{B}_{4(\downarrow)} \otimes \psi^{B}_{5(\downarrow)})$. Here, for each of the one-particle functions $\psi^{B}_{1(\uparrow)}$, $\psi^{B}_{2(\uparrow)}$, $\psi^{B}_{4(\downarrow)}$ and $\psi^{B}_{5(\downarrow)}$ we employ an s-type orbital, whereas for the one-particle function $\psi^{B}_{3(\uparrow)}$ we employ a p-type orbital. Moreover, we use a value of 1 for the parameter c and a value of 7 for the parameter L. In the case of the C atom with $M_S = 0$, we apply an $\mathcal{L}^2$-normalized partially antisymmetric function of the form $\tilde{\Psi}^{HF}_{(6,0)} = C\mathfrak{A}^{(6,0)}(\psi^{C}_{1(\uparrow)} \otimes \psi^{C}_{2(\uparrow)} \otimes \psi^{C}_{3(\uparrow)} \otimes \psi^{C}_{4(\downarrow)} \otimes \psi^{C}_{5(\downarrow)} \otimes \psi^{C}_{6(\downarrow)})$ for the rank-1 approximation in (4.36). We set $\psi^{Be}_{1(\uparrow)} = \psi^{Be}_{4(\downarrow)} = o_1^{(s)}$, $\psi^{Be}_{2(\uparrow)} = \psi^{Be}_{5(\downarrow)} = o_2^{(s)}$ and $\psi^{Be}_{3(\uparrow)} = \psi^{Be}_{6(\downarrow)} = o_1^{(p)}$. Here, o_1^s and o_2^s denote s-type orbitals and o_1^p denotes a p-type orbital. Moreover, we use a value of 1 for the parameter c and a value of 7 for the parameter L.

We give our numerical results according to the Li, Be, B and C atoms in Table 5.9, Table 5.10, Table 5.11 and Table 5.12, respectively. Also, our results are compared to that obtained by other methods in Table 5.13[2] and are summarized in Figure 5.8.

In the case of the Li atom, we see that chemical accuracy is reached and a proportion of the correlation energies E_c of more than 97 % is obtained. In the case of the Be atom we reach an error smaller than 10^{-2} hartree. Note that for the Li and the Be atom the partial energy terms E_κ^{zero}, E_κ^{one} and $E_\kappa^{two\uparrow\downarrow}$ are of the largest size within the particle-wise decomposition. In the case of the B and the C atom our results seem to be in the pre-asymptotic range. However, we still obtain proportions of the correlation energies of more than 87 % and 70 %, respectively. Moreover, for all considered atoms, except for the C atom, the error of our results is of smaller size than the error related to the results obtained by the variational Monte Carlo method [21] based on a single-determinant Jastrow-Slater trial wave function. In particular, our results are in the range of the results computed by a diffusion Monte Carlo method [21] based on a single-determinant Jastrow-Slater trial wave function.

---

[2] Note that in the case of the B atom the value for $\tilde{E}^{HF}_{tot}$ is slightly less than the value for the Hartree-Fock limit E^{HF}_{tot} taken from literature [38]. This is due to that we employ an unrestricted open-shell Hartree-Fock approximation.

Table 5.9: Numerical results of Algorithm 4.1 for the Li atom.

κ	M_κ	$\tilde{E}_\kappa^{(1)}$	e_κ	s_κ^M	s_κ^ε
1	610	-7.471645	6.42_{-3}		
4	634	-7.473208	4.85_{-3}		
6	850	-7.474914	3.15_{-3}		
7	1,150	-7.475572	2.49_{-3}	-0.78	0.34
8	2,350	-7.476317	1.74_{-3}	-0.50	0.51
9	8,179	-7.476838	1.22_{-3}	-0.28	0.51
10	22,203	-7.477022	1.04_{-3}	-0.16	0.23

κ	E_κ^{zero}	E_κ^{one}	$E_\kappa^{two\uparrow\downarrow}$	$E_\kappa^{two\uparrow\uparrow}$	E_κ^{three}
1	-6.83	-1.24	0.58	1.88_{-2}	2.15_{-3}
4	-6.13	-2.65	1.29	2.36_{-2}	-2.41_{-3}
6	-6.21	-2.49	1.20	2.10_{-2}	5.46_{-4}
7	-5.54	-3.81	1.85	2.08_{-2}	5.91_{-4}
8	-5.84	-3.22	1.58	3.40_{-3}	4.87_{-4}
9	-5.83	-3.21	1.56	6.59_{-3}	7.63_{-4}
10	-5.66	-3.53	1.71	7.55_{-3}	1.03_{-3}

κ	M_κ	M_κ^{one}	$M_\kappa^{two\uparrow\downarrow}$	$M_\kappa^{two\uparrow\uparrow}$	M_κ^{three}
1	610	253	190	108	58
4	634	253	214	108	58
6	850	313	370	108	58
7	1,150	313	670	108	58
8	2,350	445	1,702	144	58
9	8,179	1,255	4,720	2,145	58
10	22,203	1,880	16,018	4,246	58

Table 5.10: Numerical results of Algorithm 4.1 for the Be atom.

κ	M_κ	$\tilde{E}_\kappa^{(1)}$	e_κ	s_κ^M	s_κ^ε
1	1,417	−14.644213	2.31_{-2}		
4	1,633	−14.650281	1.71_{-2}		
6	3,553	−14.654086	1.33_{-2}		
7	6,667	−14.656040	1.13_{-2}	−0.25	0.23
8	10,067	−14.657765	9.59_{-3}	−0.40	0.24
9	24,775	−14.659783	7.58_{-3}	−0.26	0.34

κ	E_κ^{zero}	E_κ^{one}	$E_\kappa^{two\uparrow\downarrow}$	$E_\kappa^{two\uparrow\uparrow}$	E_κ^{three}	E_κ^{other}
1	−9.83	−8.00	3.13	5.41_{-2}	-5.36_{-4}	1.43_{-6}
4	−5.84	−15.79	6.93	5.13_{-2}	-5.15_{-5}	1.91_{-6}
6	3.08	−34.35	16.57	4.99_{-2}	-5.95_{-5}	2.25_{-6}
7	2.18	−32.68	15.79	4.96_{-2}	-1.33_{-4}	2.29_{-6}
8	−8.25	−13.45	6.99	4.96_{-2}	-1.36_{-4}	2.26_{-6}
9	−12.31	−6.06	3.66	4.88_{-2}	-1.79_{-5}	2.12_{-6}

κ	M_κ	M_κ^{one}	$M_\kappa^{two\uparrow\downarrow}$	$M_\kappa^{two\uparrow\uparrow}$	M_κ^{three}	M_κ^{other}
1	1,417	524	562	264	64	2
4	1,633	584	718	264	64	2
6	3,553	800	2,422	264	64	2
7	6,667	1,100	5,236	264	64	2
8	10,067	1,976	7,760	264	64	2
9	24,775	2,780	21,664	264	64	2

Table 5.11: Numerical results of Algorithm 4.1 for the B atom.

κ	M_κ	$\tilde{E}_\kappa^{(1)}$	e_κ	s_κ^M	s_κ^ε
1	2,637	−24.620106	3.38_{-2}		
3	2,661	−24.620191	3.37_{-2}		
4	3,119	−24.624262	2.96_{-2}	−0.81	0.19
5	3,415	−24.628132	2.58_{-2}	−1.54	0.20
6	6,197	−24.632901	2.10_{-2}	−0.34	0.30
7	17,137	−24.637678	1.62_{-2}	−0.25	0.37

κ	E_κ^{zero}	E_κ^{one}	$E_\kappa^{two\uparrow\downarrow}$	$E_\kappa^{two\uparrow\uparrow}$	E_κ^{three}	E_κ^{other}
1	−9.68	−25.55	10.32	1.43_{-1}	1.44_{-1}	3.75_{-5}
3	−9.00	−26.24	10.33	1.44_{-1}	1.43_{-1}	3.76_{-5}
4	−10.65	−21.99	7.74	8.87_{-2}	1.93_{-1}	3.67_{-5}
5	−8.99	−25.10	9.14	2.71_{-1}	5.17_{-2}	3.81_{-5}
6	−6.72	−30.47	12.24	2.67_{-1}	5.41_{-2}	3.64_{-5}
7	−0.55	−40.52	14.18	2.20_0	5.57_{-2}	4.03_{-6}

κ	M_κ	M_κ^{one}	$M_\kappa^{two\uparrow\downarrow}$	$M_\kappa^{two\uparrow\uparrow}$	M_κ^{three}	M_κ^{other}
1	2,637	590	610	646	788	2
3	2,661	614	610	646	788	2
4	3,119	650	1,032	646	788	2
5	3,415	767	1,211	646	788	2
6	6,197	1,001	3,759	646	788	2
7	17,137	1,813	13,397	1,136	788	2

Table 5.12: Numerical results of Algorithm 4.1 for the C atom.

κ	M_κ	$E_\kappa^{(1)}$	e_κ	s_κ^M	s_κ^ε
1	4,258	−37.767724	7.73_{-2}		
3	4,452	−37.771020	7.40_{-2}		
4	4,662	−37.776946	6.81_{-2}	−1.81	0.12
5	5,612	−37.782707	6.23_{-2}	−0.48	0.13
6	12,197	−37.791802	5.32_{-2}	−0.20	0.23
7	23,875	−37.800198	4.48_{-2}	−0.26	0.25

κ	E_κ^{zero}	E_κ^{one}	$E_\kappa^{two\uparrow\downarrow}$	$E_\kappa^{two\uparrow\uparrow}$	E_κ^{three}
1	−6.84	−56.66	25.45	2.88_{-1}	-5.49_{-3}
3	−9.22	−48.14	19.31	2.26_{-1}	5.25_{-2}
4	−17.62	−29.03	8.55	3.42_{-1}	-1.69_{-2}
5	−18.64	−30.53	11.05	1.44_{-2}	3.20_{-1}
6	−40.97	8.26	−9.19	3.82_{0}	2.91_{-1}
7	−56.21	10.27	0.99	6.59_{0}	5.58_{-1}

κ	M_κ	M_κ^{one}	$M_\kappa^{two\uparrow\downarrow}$	$M_\kappa^{two\uparrow\uparrow}$	M_κ^{three}
1	4,258	716	915	1,022	1,604
3	4,452	764	1,061	1,022	1,604
4	4,662	790	1,245	1,022	1,604
5	5,612	874	2,111	1,022	1,604
6	12,197	1,636	6,791	2,165	1,604
7	23,875	3,351	15,507	3,412	1,604

Table 5.13: Comparison of numerical results for the atoms Li, Be, B and C. The values for estimated *exact* total energies E_{tot} are taken from [29] and the values for the Hartree-Fock limits E_{tot}^{HF} are given in [38]. The total energy values E_{tot}^{VMC} and E_{tot}^{DMC} computed by variational Monte Carlo (VMC) and diffusion Monte Carlo (DMC) based on a single-determinant Jastrow-Slater trial wave function are taken from [21]. The energies values $E_{tot}^{CCSD(T)}$ and E_{tot}^{FCI} are obtained by a CCSD(T) and a Full-CI computation, respectively, using the PSI3 [36] package with the cc-pVTZ basis set [47].

	Li	Be	B	C
$\tilde{E}_{tot}$	−7.47702	−14.65978	−24.63768	−37.80020
E_{tot}	−7.47806	−14.66736	−24.65391	−37.84500
$\tilde{E}_{tot}^{HF}$	−7.43271	−14.57296	−24.52921	−37.65886
E_{tot}^{HF}	−7.43273	−14.57302	−24.52906	−37.68862
$E_c[\%]$	97.58	90.94	87.00	71.35
E_{tot}^{VMC}	−7.47683	−14.63110	−24.60562	−37.81471
E_{tot}^{DMC}	−7.47802	−14.65717	−24.63978	−37.82951
$E_{tot}^{CCSD(T)}$	−7.44607	−14.62379	−24.60538	−37.73586
E_{tot}^{FCI}	−7.44607	−14.62381	−24.60582	−37.74080

5 Numerical Experiments

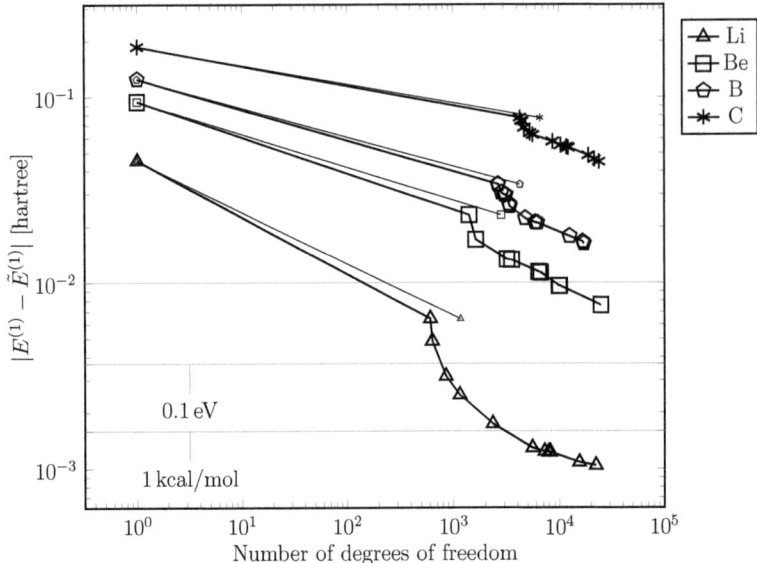

(a) Error versus number of degrees of freedom, where each step 3 in Algorithm 4.1 is taken into account; compare (4.65). The small marker denotes the initial system $\mathcal{B}^{[0]}_{three} \cup \mathcal{B}^{[0]}_{other}$ which is used in the coarsening step to obtain $\mathcal{B}^{[0]}$; compare (4.63).

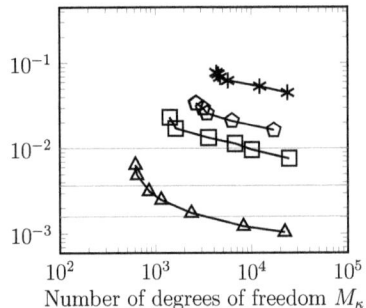 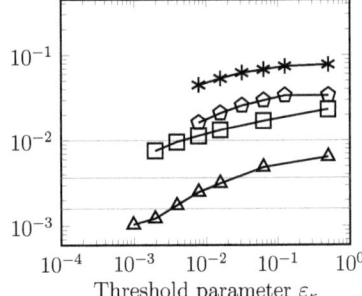

(b) Error versus the number of degrees of freedom, where each $|E^{(1)}_\kappa - \tilde{E}^{(1)}_\kappa|$ and M_κ associated with step 5 in Algorithm 4.1 is taken into account.

(c) Error versus the threshold parameter, where each $|E^{(1)}_\kappa - \tilde{E}^{(1)}_\kappa|$ and ε_κ associated with step 5 in Algorithm 4.1 is taken into account.

Figure 5.8: Error of the approximately computed energy of the ground-state of the Li, Be, B and C atom, respectively.

5.4 LiH, BeH, BH and Li$_2$ diatomic molecules

In this section we apply our new method to diatomic molecules with up to six electrons, i.e. the LiH molecule, the BeH molecule, the BH molecule and the Li$_2$ dimer.

The technical details are given in the following: We study the LiH molecule for $M_S = 0$ and a value of $R_{\text{LiH}} = 3.015$ bohr for the bond distance, i.e. $\mathbf{R}_1 = \mathbf{R}_{\text{Li}} = \mathbf{0}$ and $\mathbf{R}_2 = \mathbf{R}_{\text{H}} = (R_{\text{LiH}}, 0, 0)^T$. For the rank-1 approximation in (4.36) we apply an $\mathcal{L}^2$-normalized partially antisymmetric function of the form $\tilde{\Psi}^{HF}_{(4,0)} = C\mathfrak{A}^{(4,0)}(\psi^{\text{LiH}}_{1(\uparrow)} \otimes \psi^{\text{LiH}}_{2(\uparrow)} \otimes \psi^{\text{LiH}}_{3(\downarrow)} \otimes \psi^{\text{LiH}}_{4(\downarrow)})$. Here, for the spin-up one-particle functions we employ $\psi^{\text{LiH}}_{1(\uparrow)} = o^{(s)}_1(\mathbf{x})$ and $\psi^{\text{LiH}}_{2(\uparrow)} = o^{(s)}_2(\mathbf{x}) + o^{(s)}_3(\mathbf{x} - \mathbf{R}_H)$, where $o^{(s)}_1$, $o^{(s)}_2$ and $o^{(s)}_3$ denote s-type orbitals centered at the origin. For the spin-down one-particle functions we set $\psi^{\text{LiH}}_{3(\downarrow)} = \psi^{\text{LiH}}_{1(\uparrow)}$ and $\psi^{\text{LiH}}_{4(\downarrow)} = \psi^{\text{LiH}}_{2(\uparrow)}$. Moreover, we use a value of $\frac{2}{R_{\text{LiH}}}$ for the parameter c and a value of 6 for the parameter L. In the case of the BeH molecule with $M_S = \pm 1/2$, we consider the case of a bond distance of $R_{\text{BeH}} = 2.537$ bohr, i.e. $\mathbf{R}_1 = \mathbf{R}_{\text{Be}} = \mathbf{0}$ and $\mathbf{R}_2 = \mathbf{R}_{\text{H}} = (R_{\text{BeH}}, 0, 0)^T$, and we again focus on the case of $M_S = 1/2$ only. For the rank-1 approximation in (4.36) we apply an $\mathcal{L}^2$-normalized partially antisymmetric function of the form $\tilde{\Psi}^{HF}_{(5,1/2)} = C\mathfrak{A}^{(5,1/2)}(\psi^{\text{BeH}}_{1(\uparrow)} \otimes \psi^{\text{BeH}}_{2(\uparrow)} \otimes \psi^{\text{BeH}}_{3(\uparrow)} \otimes \psi^{\text{BeH}}_{4(\downarrow)} \otimes \psi^{\text{BeH}}_{5(\downarrow)})$. For the spin-up one-particle functions we set $\psi^{\text{BeH}}_{1(\uparrow)} = o^{(s)}_1(\mathbf{x})$, $\psi^{\text{BeH}}_{2(\uparrow)} = o^{(s)}_2(\mathbf{x})$ and $\psi^{\text{LiH}}_{3(\uparrow)} = o^{(s)}_3(\mathbf{x}) + o^{(s)}_4(\mathbf{x} - \mathbf{R}_H)$, whereas for the spin-down one-particle functions we set $\psi^{\text{BeH}}_{4(\downarrow)} = o^{(s)}_5(\mathbf{x})$ and $\psi^{\text{BeH}}_{5(\downarrow)} = o^{(s)}_6(\mathbf{x}) + o^{(s)}_7(\mathbf{x} - \mathbf{R}_H)$. Here, $o^{(s)}_1$, $o^{(s)}_2$, $o^{(s)}_3$, $o^{(s)}_4$, $o^{(s)}_5$, $o^{(s)}_6$ and $o^{(s)}_7$ are s-type orbitals centered at the origin. We apply a value of $\frac{2}{R_{\text{BeH}}}$ for the parameter c and a value of 7 for the parameter L. For the BH molecule with $M_S = 0$ we employ a value of $R_{\text{BH}} = 2.329$ bohr for the bond distance, i.e. $\mathbf{R}_1 = \mathbf{R}_B = \mathbf{0}$ and $\mathbf{R}_2 = \mathbf{R}_H = (R_{\text{BH}}, 0, 0)^T$. For the rank-1 approximation in (4.36) we apply an $\mathcal{L}^2$-normalized partially antisymmetric function of the form $\tilde{\Psi}^{HF}_{(6,0)} = C\mathfrak{A}^{(6,0)}(\psi^{\text{BH}}_{1(\uparrow)} \otimes \psi^{\text{BH}}_{2(\uparrow)} \otimes \psi^{\text{BH}}_{3(\uparrow)} \otimes \psi^{\text{BH}}_{4(\downarrow)} \otimes \psi^{\text{BH}}_{5(\downarrow)} \otimes \psi^{\text{BH}}_{6(\downarrow)})$. For the spin-up one-particle functions we employ $\psi^{\text{BH}}_{1(\uparrow)} = o^{(s)}_1(\mathbf{x})$, $\psi^{\text{BH}}_{2(\uparrow)} = o^{(s)}_2(\mathbf{x})$, and $\psi^{\text{BH}}_{3(\uparrow)} = o^{(p)}_3(\mathbf{x}) + o^{(s)}_4(\mathbf{x} - \mathbf{R}_H)$. Here, $o^{(s)}_1$, $o^{(s)}_2$ and $o^{(s)}_4$ denote s-type orbitals centered at the origin, whereas $o^{(p)}_3$ denotes a p-type orbital centered at the origin. For the spin-down one-particle functions we set $\psi^{\text{BH}}_{4(\downarrow)} = \psi^{\text{BH}}_{1(\uparrow)}$, $\psi^{\text{BH}}_{5(\downarrow)} = \psi^{\text{BH}}_{2(\uparrow)}$ and $\psi^{\text{BH}}_{6(\downarrow)} = \psi^{\text{BH}}_{3(\uparrow)}$. Furthermore, we use a value of $\frac{2}{R_{\text{BH}}}$ for the parameter c and a value of 7 for the parameter L. Finally, in the case of the Li$_2$ molecule with $M_S = 0$, we employ for the bond distance a value of $R_{\text{LiLi}} = 5.051$ bohr, i.e. $\mathbf{R}_1 = \mathbf{0}$ and $\mathbf{R}_2 = (R_{\text{LiLi}}, 0, 0)^T$. For the rank-1 approximation in (4.36) we apply an $\mathcal{L}^2$-normalized partially antisymmetric function of the form $\tilde{\Psi}^{HF}_{(6,0)} = C\mathfrak{A}^{(6,0)}(\psi^{\text{Li}_2}_{1(\uparrow)} \otimes \psi^{\text{Li}_2}_{2(\uparrow)} \otimes \psi^{\text{Li}_2}_{3(\uparrow)} \otimes \psi^{\text{Li}_2}_{4(\downarrow)} \otimes \psi^{\text{Li}_2}_{5(\downarrow)} \otimes \psi^{\text{Li}_2}_{6(\downarrow)})$. Here, for the spin-up one-particle functions we employ $\psi^{\text{Li}_2}_{1(\uparrow)} = o^{(s)}_1(\mathbf{x})$, $\psi^{\text{Li}_2}_{2(\uparrow)} = o^{(s)}_2(\mathbf{x})$ and $\psi^{\text{Li}_2}_{3(\uparrow)} = o^{(s)}_3(\mathbf{x} - \mathbf{R}_2)$. For the spin-down one-particle functions we set $\psi^{\text{Li}_2}_{4(\downarrow)} = o^{(s)}_1(\mathbf{x} - \mathbf{R}_2)$, $\psi^{\text{Li}_2}_{5(\downarrow)} = o^{(s)}_2(\mathbf{x} - \mathbf{R}_2)$ and $\psi^{\text{Li}_2}_{6(\downarrow)} = o^{(s)}_3(\mathbf{x})$. Here, $o^{(s)}_1$, $o^{(s)}_2$ and $o^{(s)}_3$ denote s-type orbitals centered at the origin. We use a value of $\frac{2}{R_{\text{LiLi}}}$ for the parameter c and a value of 7 for the parameter L.

Our numerical results according the LiH, BeH, BH and Li$_2$ diatomic molecules are given in Table 5.14, Table 5.15, Table 5.16 and Table 5.17. Additionally, we depict the results in Figure 5.9 and compare them with those computed by other methods in Table 5.18.

In the case of the LiH atom we observe an approximation error smaller than a value

5 Numerical Experiments

of 10^{-2} hartree and a proportion of the correlation energies E_c of more than 88 %. We further see that for the diatomic molecules LiH, BeH and Li$_2$ the partial energy terms E_κ^{zero}, E_κ^{one} and $E_\kappa^{two\uparrow\downarrow}$ are of the largest size within the particle-wise decomposition. In the case of the molecules BeH, BH and Li$_2$ our results seem to be in the pre-asymptotic range. However, we still obtain proportions of the correlation energies of more than 71 %, 82 % and 70 %, respectively. Moreover, for all systems except the Li$_2$ molecule the error of our results is of smaller size than the error related to the results obtained by a variational Monte Carlo method [127, 177] based on a single-determinant Jastrow-Slater trial wave function. Furthermore, the approximation errors of our results are of smaller size than those obtained by CCSD(T) and Full-CI computations using the PSI3 package [36] with the cc-pVQZ basis set [47].[3]

Table 5.14: Numerical results of Algorithm 4.1 for the LiH molecule.

κ	M_κ	$\tilde{E}_\kappa^{(1)}$	e_κ	s_κ^M	s_κ^ε
1	2,823	−9.039973	2.56_{-2}		
4	3,345	−9.047351	1.82_{-2}		
5	4,937	−9.052533	1.30_{-2}	−0.86	0.48
6	17,867	−9.055869	9.71_{-3}	−0.23	0.43

κ	E_κ^{zero}	E_κ^{one}	$E_\kappa^{two\uparrow\downarrow}$	$E_\kappa^{two\uparrow\uparrow}$	E_κ^{three}	E_κ^{other}
1	−2.45	−9.62	3.00	1.77_{-2}	8.73_{-3}	-4.30_{-4}
4	−3.03	−5.68	−0.36	1.82_{-2}	7.31_{-3}	-4.86_{-4}
5	0.79	−10.38	0.52	1.88_{-2}	6.28_{-3}	-5.16_{-4}
6	6.23	−17.07	1.77	1.87_{-2}	6.30_{-3}	-5.20_{-4}

κ	M_κ	M_κ^{one}	$M_\kappa^{two\uparrow\downarrow}$	$M_\kappa^{two\uparrow\uparrow}$	M_κ^{three}	M_κ^{other}
1	2,823	922	598	574	722	6
4	3,345	1,010	1,032	574	722	6
5	4,937	1,058	2,576	574	722	6
6	17,867	1,744	14,820	574	722	6

---

[3] Note that in Table 5.18 the Full-CI result for the Li$_2$ molecule is missing due to physical memory limitations.

Table 5.15: Numerical results of Algorithm 4.1 for the BeH molecule.

κ	M_κ	$\tilde{E}_\kappa^{(1)}$	e_κ	s_κ^M	s_κ^ε
1	4,132	-16.776760	4.67_{-2}		
2	4,144	-16.778034	4.54_{-2}		
4	4,596	-16.786366	3.71_{-2}		
5	7,839	-16.793625	2.98_{-2}	-0.41	0.31
6	23,564	-16.796828	2.66_{-2}	-0.10	0.16

κ	E_κ^{zero}	E_κ^{one}	$E_\kappa^{two\uparrow\downarrow}$	$E_\kappa^{two\uparrow\uparrow}$	E_κ^{three}	E_κ^{other}
1	-5.12	-16.94	5.10	-6.13_{-2}	2.43_{-1}	1.38_{-5}
2	-4.12	-17.77	4.93	-5.24_{-2}	2.41_{-1}	1.37_{-5}
4	-0.01	-22.05	5.07	1.25_{-1}	7.76_{-2}	1.11_{-5}
5	-0.02	-18.22	1.23	1.71_{-1}	3.99_{-2}	9.97_{-6}
6	-1.72	-16.28	1.00	1.41_{-1}	6.39_{-2}	1.01_{-5}

κ	M_κ	M_κ^{one}	$M_\kappa^{two\uparrow\downarrow}$	$M_\kappa^{two\uparrow\uparrow}$	M_κ^{three}	M_κ^{other}
1	4,132	995	848	946	1,340	2
2	4,144	1,007	848	946	1,340	2
4	4,596	1,215	1,092	946	1,340	2
5	7,839	1,733	3,817	946	1,340	2
6	23,564	3,441	17,834	946	1,340	2

Table 5.16: Numerical results of Algorithm 4.1 for the BH molecule.

κ	M_κ	$\tilde{E}_\kappa^{(1)}$	e_κ	s_κ^M	s_κ^ε
1	6,594	−27.375788	5.90_{-2}		
2	6,642	−27.380733	5.40_{-2}	−12.03	0.13
3	6,690	−27.380921	5.38_{-2}	−0.49	0.01
4	6,947	−27.385515	4.92_{-2}	−2.37	0.13
5	8,048	−27.391574	4.32_{-2}	−0.89	0.19
6	15,907	−27.403709	3.10_{-2}	−0.48	0.48
7	28,125	−27.407739	2.70_{-2}	−0.24	0.20

κ	E_κ^{zero}	E_κ^{one}	$E_\kappa^{two\uparrow\downarrow}$	$E_\kappa^{two\uparrow\uparrow}$	E_κ^{three}	E_κ^{other}
1	3.38	−54.35	23.41	1.17_{-1}	6.57_{-2}	-3.30_{-6}
2	4.98	−54.70	22.17	2.67_{-2}	1.44_{-1}	-3.38_{-6}
3	5.55	−55.27	22.19	-3.70_{-3}	1.47_{-1}	-3.37_{-6}
4	8.32	−54.50	18.66	-4.03_{-2}	1.74_{-1}	-3.53_{-6}
5	14.97	−56.98	14.42	6.85_{-2}	1.29_{-1}	-3.78_{-6}
6	8.72	−44.23	7.90	-1.20_{-2}	2.25_{-1}	-3.78_{-6}
7	1.78	−36.27	5.49	1.05_{0}	5.48_{-1}	-3.55_{-6}

κ	M_κ	M_κ^{one}	$M_\kappa^{two\uparrow\downarrow}$	$M_\kappa^{two\uparrow\uparrow}$	M_κ^{three}	M_κ^{other}
1	6,594	1,236	1,291	1,544	2,520	2
2	6,642	1,260	1,315	1,544	2,520	2
3	6,690	1,308	1,315	1,544	2,520	2
4	6,947	1,380	1,500	1,544	2,520	2
5	8,048	1,759	2,222	1,544	2,520	2
6	15,907	3,590	8,250	1,544	2,520	2
7	28,125	4,769	19,097	1,676	2,580	2

Table 5.17: Numerical results of Algorithm 4.1 for the Li_2 molecule.

κ	M_κ	$\tilde{E}_\kappa^{(1)}$	e_κ	s_κ^M	s_κ^ε
1	3,846	−16.733156	4.39_{-2}		
2	3,878	−16.735794	4.12_{-2}	−7.40	0.09
3	4,075	−16.739737	3.73_{-2}	−2.03	0.15
4	4,460	−16.740562	3.64_{-2}	−0.25	0.03
5	5,365	−16.741125	3.59_{-2}	−0.08	0.02
6	7,404	−16.742013	3.50_{-2}	−0.08	0.04
7	12,648	−16.748830	2.82_{-2}	−0.40	0.31
8	24,277	−16.749711	2.73_{-2}	−0.05	0.05

κ	E_κ^{zero}	E_κ^{one}	$E_\kappa^{two\uparrow\downarrow}$	$E_\kappa^{two\uparrow\uparrow}$	E_κ^{three}
1	−16.95	−3.61	3.78	3.53_{-2}	9.01_{-3}
2	−9.72	−12.45	5.38	5.87_{-2}	-1.64_{-3}
3	−8.64	−10.37	2.21	5.85_{-2}	-3.44_{-3}
4	−11.22	−3.07	−2.51	5.83_{-2}	-2.70_{-3}
5	−8.90	−5.53	−2.37	5.90_{-2}	-2.49_{-3}
6	−7.11	−7.06	−2.63	5.71_{-2}	4.85_{-3}
7	−6.32	−10.20	−0.29	5.70_{-2}	5.31_{-3}
8	−6.05	−10.55	−0.21	5.65_{-2}	3.24_{-3}

κ	M_κ	M_κ^{one}	$M_\kappa^{two\uparrow\downarrow}$	$M_\kappa^{two\uparrow\uparrow}$	M_κ^{three}
1	3,846	810	1,243	920	872
2	3,878	842	1,243	920	872
3	4,075	842	1,440	920	872
4	4,460	842	1,825	920	872
5	5,365	842	2,730	920	872
6	7,404	1,082	4,529	920	872
7	12,648	1,386	9,469	920	872
8	24,277	2,000	20,484	920	872

Table 5.18: Comparison of numerical results for the LiH, the BeH, the BH and the Li$_2$ molecules. The energies E_{tot} are taken from [127], except for LiH and Li$_2$ which are taken from [28] and [49], respectively. The Hartree-Fock limits E_{tot}^{HF} are taken from [127], except for Li$_2$ which is taken from [49]. The total energy values E_{tot}^{VMC} and E_{tot}^{DMC} computed by variational Monte Carlo (VMC) and diffusion Monte Carlo (DMC) based on a single-determinant Jastrow-Slater trial wave function are taken from [127] and for Li$_2$ from [177]. The energies $E_{tot}^{CCSD(T)}$ and E_{tot}^{FCI} are obtained by a CCSD(T) and a Full-CI computation, respectively, using the PSI3 [36] package with the cc-pVQZ basis set [47].

	LiH	BeH	BH	Li$_2$	
$\tilde{E}_{tot}$	-8.06084	-15.22016	-25.26089	-14.96810	
E_{tot}	-8.07055	-15.24680	-25.28790	-14.99540	
$\tilde{E}_{tot}^{HF}$	-7.97054	-15.04328	-25.08382	-14.86166	
E_{tot}^{HF}	-7.98735	-15.15318	-25.13195	-14.87152	
$E_c[\%]$	88.05	71.55	82.68	77.97	
E_{tot}^{VMC}	-8.04593	-15.21210	-25.21220	-14.98255	
E_{tot}^{DMC}	-8.07012	-15.24062	-25.27595	-14.99167	
$E_{tot}^{CCSD(T)}$	-8.04267	-15.12930	-25.04420	-14.94112	cc-pVQZ
E_{tot}^{FCI}	-8.04269	-15.12712	-25.04123		cc-pVQZ

5.4 LiH, BeH, BH and Li_2 diatomic molecules

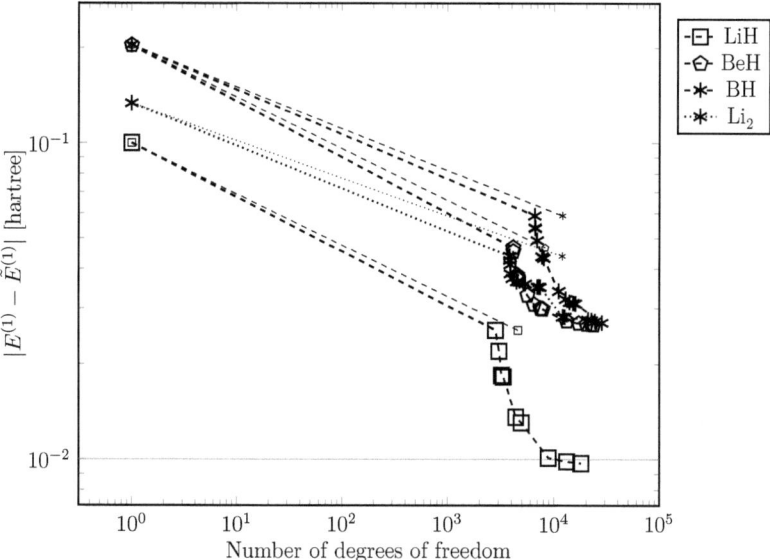

(a) Error versus number of degrees of freedom, where each step 3 in Algorithm 4.1 is taken into account; compare (4.65). The small marker denotes the initial system $\mathcal{B}^{[0]}_{three} \cup \mathcal{B}^{[0]}_{other}$ which is used in the coarsening step to obtain $\mathcal{B}^{[0]}$; compare (4.63).

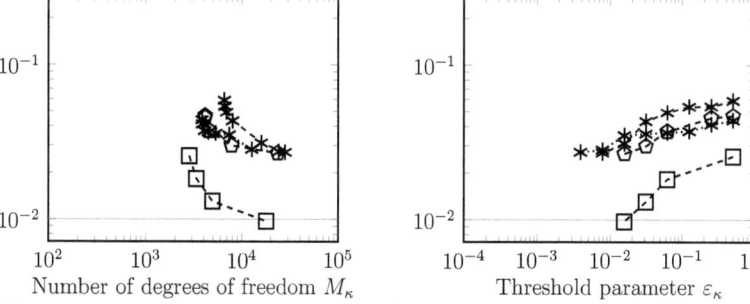

(b) Error versus the number of degrees of freedom, where each $|E^{(1)}_\kappa - \tilde{E}^{(1)}_\kappa|$ and M_κ associated with step 5 in Algorithm 4.1 is taken into account.

(c) Error versus the threshold parameter, where each $|E^{(1)}_\kappa - \tilde{E}^{(1)}_\kappa|$ and ε_κ associated with step 5 in Algorithm 4.1 is taken into account.

Figure 5.9: Error of the approximately computed energy of the ground-state of the LiH, BeH, BH and Li_2 molecules.

5.5 Summary

We applied our new method introduced in Chapter 4 to small molecular systems with up to six electrons. The results of the numerical experiments suggest that our new method is indeed convergent and that the measured rates are in the expected range. In particular, the results demonstrate that our new method allows to efficiently describe electron-nuclei and electron-electron cusps. Nevertheless, we see a dependence of the involved constants on the molecular system size.

For a better overview, we summarized the used parameters and our numerical results for the ground-states energies of the atoms and molecules according to Section 5.1, Section 5.3 and Section 5.4 in Table 5.19 and Figure 5.10. In addition, we depicted these results with respect to the proportion of the correlation energy E_c in Figure 5.11.

For the two- and three-electron systems chemical accuracy is reached. For four-electron systems we obtain an approximation error of a size smaller than ten milli-hartree. In particular, we obtain in the case of two-electron, three-electron and four-electron systems systems proportions greater than 98 %, 96 % and 88 %, respectively. In the case of the five- and six-electron molecular systems our results seem to be still in the pre-asymptotic range. Note here that the involved number of degrees of freedom is limited by physical memory limitations. However, for the five- and six-electron systems proportions of the correlation energies in the range of 71 % to 87 % are achieved.

For all studied systems, except for the H_2 molecule, the size of the approximation error for our new method to compute to the total energy is less than the error obtained by a CCSD(T) and a Full-CI computation with the help of the PSI3 package using certain LCAO basis sets. In particular, for all systems considered in Section 5.3 and Section 5.4, except for the six-electron molecules, the size of the approximation error of our new method is less than the error obtained by variational Monte Carlo methods which employ a single-determinant Jastrow-Slater trial wave function. Moreover, for all molecular systems in Section 5.3 and Section 5.4 our results are in the range of the results computed by diffusion Monte Carlo methods which are based on a single-determinant Jastrow-Slater trial wave function.

Table 5.19: Parameters for the frame $\mathfrak{b}_{\sigma,c}$ and its subset $\mathfrak{b}_{\sigma,c}^{(L)}$; compare (4.29) and (4.51). For all systems we set $\sigma = 1$. For all diatomic systems we set $c = \frac{2}{R}$. The respective bond distances R are given in bohr.

	He	(^3S)He	$(^4P^0)$Li	Li	Be	B	C
c	$\frac{1}{2}$	$\frac{1}{4}$	$\frac{1}{4}$	1	$\frac{1}{4}$	1	1
L	7	7	7	6	8	7	7

	H_2	$(b^3\Sigma_u^+)H_2$	He_2^+	LiH	BeH	BH	Li_2
R	1.4	2.0	2.042	3.015	2.537	2.329	5.051
L	5	5	6	6	7	7	7

5.5 Summary

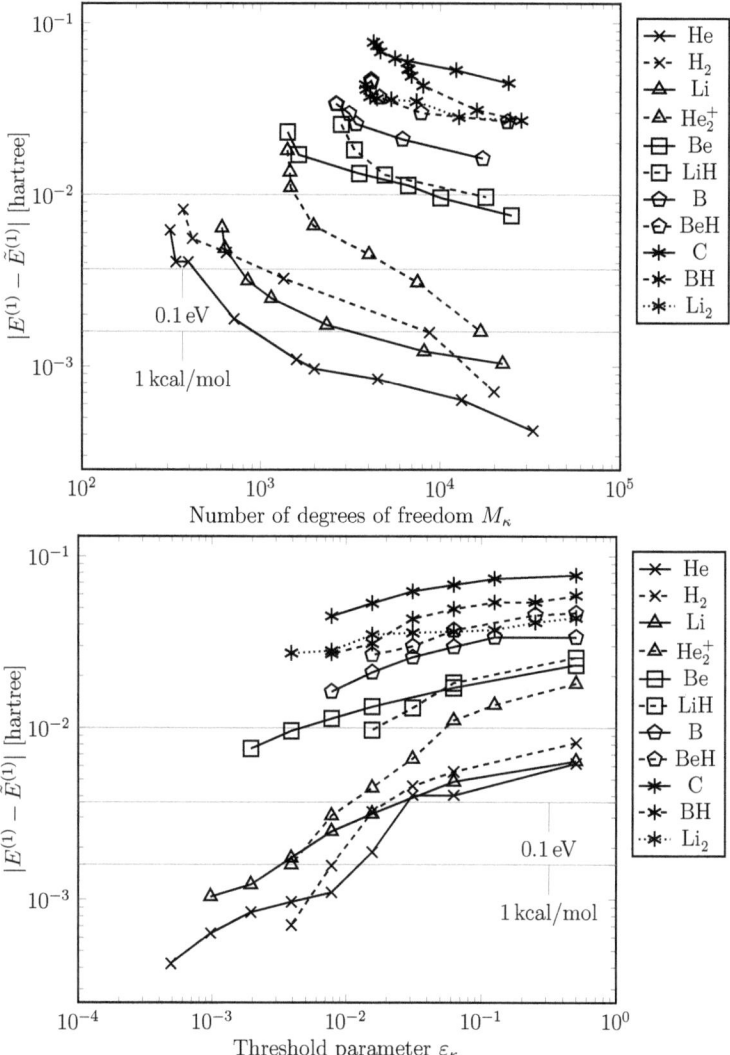

Figure 5.10: Error of the approximately computed ground-state energy of the studied atoms and molecules.

5 Numerical Experiments

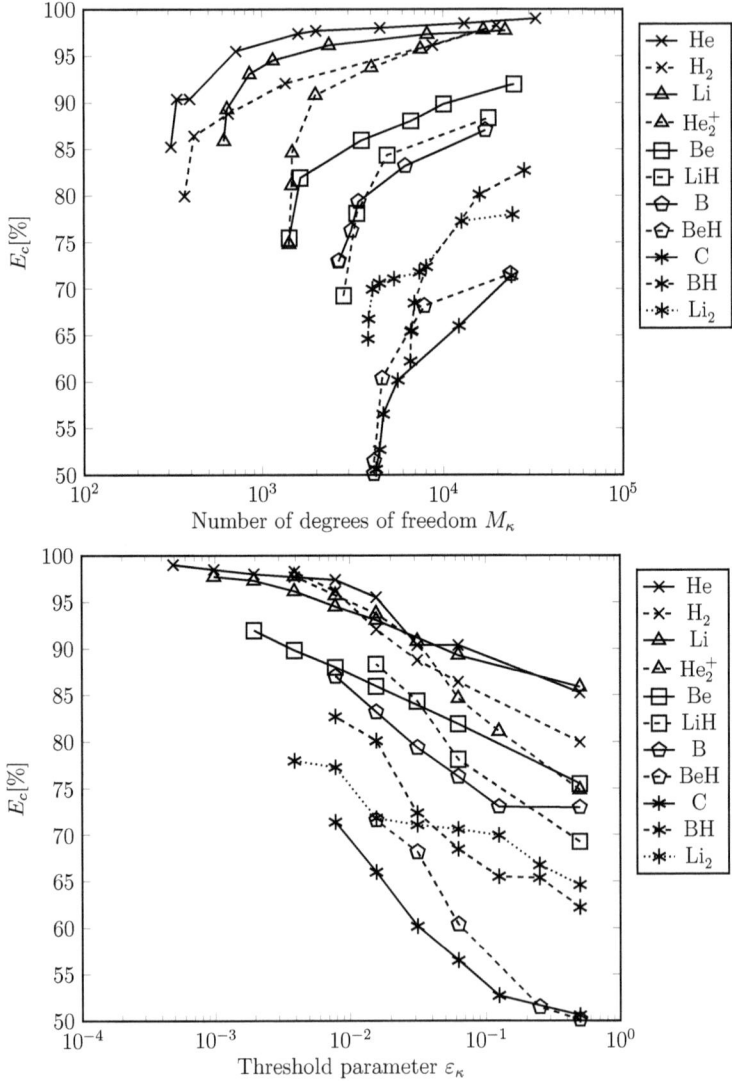

Figure 5.11: Proportion of correlation energies of the approximately computed ground-state energy of the studied atoms and molecules; see (5.1).

6 Conclusions

In this thesis we introduced and studied new tensor product multiscale many-particle spaces with finite-order weights and applied them for the numerical treatment of the electronic Schrödinger equation. These spaces are constructed from a particle-wise subspace splitting of the N-particle space. In particular, this construction provides a systematic improvement of a nonlinear rank-1 approximation by its combination with a tensor product multiscale approximation scheme. Moreover, it includes the well-known configuration interaction spaces from quantum chemistry as a special case. In contrast to many widely used (nonlinear and/or non-variational) methods in quantum chemistry, our approach provides convergence with guaranteed approximation rates. We demonstrated these rates numerically for atoms and diatomic molecules with up to six electrons. Note that to our knowledge this is the first time that systems with more than two electrons were successfully treated by the application of tensor product multiscale bases in the framework of ab initio methods except for HF and DFT methods.

In order to realize our new approach, we introduced and studied tensor product multiscale many-particle spaces. To this end, we first generalized the optimized sparse grid approximation spaces for one-dimensional particles with periodic boundary conditions to the case of N D-dimensional particles with certain decay conditions. For functions with certain smoothness and decay assumptions, we showed estimates for the approximation and complexity orders. We further generalized the sparse grid approximation to the case of antisymmetry by replacing the conventional product with the outer product which involves the Slater determinant construction. Additional conditions on the indices of the multivariate basis were imposed which reflect the Pauli principle. We thus obtained a true basis for antisymmetric sparse grid spaces with a substantially reduced amount of degrees of freedom and derived associated complexities and approximation properties.

Using our theoretical results as well as known regularity and decay properties for the eigenfunctions in the discrete spectrum of the electronic Schrödinger operator, we found that, up to logarithmic terms, the convergence rate is independent of the number of electrons N and in particular almost the same as in the two-electron case. However, the constants involved in the approximation and complexity order estimates could depend exponentially on the number of particles, which makes the antisymmetric general sparse grid spaces practical only for a moderate number of electrons.

In order to treat higher numbers of particles, we considered tensor product many-particle spaces with finite-order weights. We introduced a general particle-wise subspace splitting for the many-particle space based on splittings of the one-particle space and the corresponding decomposition of many-particle functions. For example, our formulation includes the ANOVA decomposition known from statistics and the HDMR decomposition known from computational chemistry. Furthermore, we explained the use of finite-order weights for a restriction of this subspace splitting and gave the corresponding complexities and approximation properties. These spaces allow to get rid

6 Conclusions

of the exponential dependence on the number of particles, since for finite-order weights of order q the problem of an approximation of an N-particle function reduces to the problem of the approximation of q-particle functions. In addition, we discussed the generalization of this particle-wise subspace splitting to antisymmetry and the relation to the configuration interaction spaces known from quantum chemistry.

We applied the Galerkin approach for the electronic Schrödinger equation using a sequence of finite-dimensional subspaces constructed with a particle-wise subspace splitting and an h-adaptive scheme for general sparse grid spaces. This splitting is based on one-particle functions resulting from a nonlinear rank-1 approximation and in particular corresponding to non-orthogonal Hartree-Fock orbitals. To allow for an efficient computation of the necessary matrix entries, we introduced multiscale Gaussian frames which can be seen as an approximation of a Meyer wavelet. In this way, all one- and two-particle operator integrals involved in the Löwdin rules can be written in terms of analytic formulae. Then, we introduced an algorithm to build a subspace sequence by refinement and expansion starting with an appropriate subspace. We presented a construction scheme for this initial space based on an a priori selection of subspaces in the framework of the particle-wise subspace splitting and of finite subsets of a multiscale Gaussian frame. Furthermore, we set up the matrices in parallel and solved the associated discrete eigenvalue problem with a parallelized locally optimal block preconditioned conjugate gradient method.

Finally, we applied our novel approach to atomic and diatomic systems with up to six electrons, which corresponds to a general linear eigenvalue problem in 18 space dimensions. We compared costs, accuracy and convergence rates. The numerical experiments demonstrated that our new numerical method indeed allows for convergence with the expected rates. The obtained absolute accuracy is only limited by physical memory requirements and in the range of the high-accuracy standard diffusion quantum Monte Carlo method based on a single-determinant Jastrow-Slater trial wave function.

Altogether, we demonstrated that our novel approach is an important step towards an efficient direct numerical treatment of many-particle problems like the electronic Schrödinger equation by the combination of nonlinear low-rank approximation and tensor product multiscale based approximation methods.

Certainly, in this thesis there remain open questions and problems concerning further theoretical results and the improvement of the practical applicability of our novel approach to quantum chemistry. The question of size consistency, for example, usually comes up in standard truncated configuration-interaction methods based on linear approximation. Some further open questions and problems are worth mentioning for future research.

Both the multiscale frame and the adaptive scheme can probably still be improved. Multi-wavelet like frames based on Hermite-Gaussian functions together with an h- and p-adaptive refinement strategy may lead to better approximation properties. Furthermore, from a theoretical point of view, a best M-term approximation requires a new, not yet existing Besov regularity theory in the setting of the electronic Schrödinger equation and hence, presents an opportunity for future research. Moreover, in the framework of the particle-wise subspace splitting, we currently choose finite-order weights to switch on subspaces in an a priori fashion. It would be interesting to extend our scheme to vary weights and thereby subspaces in an adaptive way (particle-wise adaptivity) similar to dimension-wise adaptive approaches for high-dimensional quadrature [62, 96]. However, this extension so far lacks theoretical justification.

Besides the generalization to the application of nonlinear approximations of a rank greater than one (cf. multi-configuration Hartree-Fock), a further promising approach, similar to the R12 and F12 methods [105], would be the employment of two-particle functions for a more efficient representation of the electron-electron cusps. In particular, the presented particle-wise decomposition based on a product of one-particle functions could be generalized to the case of a product of two-particle functions [169] according to the extended geminal model from quantum chemistry [158]. Also, an adaptive scheme which applies multiscale frames including ridgelet-like two-particle functions [46] could be a promising tool here. If, for example, the two-particle functions are built from Gaussians, all necessary operator integrals can also be given in terms of analytic formulae.

The approximation method employed in this thesis seeks to minimize the error of the ground state energy. However, in quantum chemistry, the most interesting results are not given in absolute values of eigenvalues but by their difference, e.g. for the computation of ionization and binding energies. Hence, the approximation and complexity estimates as well as the implemented scheme may be adapted to these cases in the future. Another interesting point is to use a regularized molecular Hamiltonian. This allows for an efficient computation of the necessary matrix entries for various multiscale frames. Then, however, the problem is to control and balance the different error terms. Nevertheless, similar to pseudopotential methods, such an approach may then be applied in the framework of the approximation of energy differences only.

We finally note that our approach is not restricted to the treatment of the electronic Schrödinger equation but can also be applied to other many-particle or high-dimensional problems.

6 Conclusions

A Function Spaces

In this thesis we mainly deal with one-particle isotropic Sobolev spaces $\mathcal{H}^r(\mathbb{R}^n)$ and anisotropic many-particle Sobolev spaces of dominated mixed smoothness $\mathcal{H}_{\text{mix}}^{t,r}((\mathbb{R}^D)^N)$. However, we refer sometimes to other spaces. For a better understanding, we give the definitions of these in the following.

A.1 Hölder spaces

Definition A.1. *A function $f : \mathbb{R}^n \to \mathbb{C}$ is Hölder continuous, when for all $\vec{x}, \vec{y} \in \mathbb{R}^n$ it holds*
$$|f(\vec{x}) - f(\vec{y})| \leq C|\vec{x} - \vec{y}|_2^\alpha$$
for $C > 0$ and $\alpha > 0$.

Definition A.2. *Let Ω an open subset of $\mathbb{R}^n$. Then, the Hölder space $C^{m,\alpha}(\Omega)$ with $m \in \mathbb{N}_0$, $\alpha \in [0,1]$ is the vector space of all functions $f : \mathbb{R}^n \to \mathbb{C}$ for which the norm*
$$\|f\|_{C^{m,\alpha}} := \|f\|_{C^m} + \max_{|\vec{\beta}|_1 = m} |D^{\vec{\beta}} f|_{C^{0,\alpha}}$$
is bounded. Here, the norm $\|\cdot\|_{C^m}$ is given by
$$\|f\|_{C^m} := \max_{|\vec{\beta}|_1 \leq m} \sup_{\vec{x} \in \Omega} |D^{\vec{\beta}} f|$$
and the semi-norm $|\cdot|_{C^{0,m}}$ is given by
$$|f|_{C^{0,\alpha}} := \sup_{\vec{x}, \vec{y} \in \Omega} \frac{|f(\vec{x}) - f(\vec{y})|}{|\vec{x} - \vec{y}|_2^\alpha}.$$

A.2 Besov and Triebel-Lizorkin spaces

According to (2.27), $\tilde{Q}_l$ denotes the domain $\tilde{Q}_l = \{\vec{k} \in \mathbb{R}^n : |\vec{k}|_\infty \leq 2^l\}$.

Definition A.3. *A family of functions $\{\tilde{\eta}_l\}_{l \in \mathbb{N}_0} \subset \mathcal{S}(\mathbb{R}^n)$ forms an (inhomogeneous) smooth dyadic partition of unity if*
$$\text{supp } \tilde{\eta}_0 \subset \tilde{Q}_1,$$
$$\text{supp } \tilde{\eta}_l \subset \tilde{Q}_{l+1} \setminus \tilde{Q}_{l-1}, l \in \mathbb{N},$$
$$\sup_{\mathbf{k} \in \mathbb{R}^n, l \in \mathbb{N}_0} 2^{l|\vec{\beta}|_1} |D^{\vec{\alpha}} \tilde{\eta}_l(\mathbf{k})| \leq C_{\vec{\alpha}} < \infty, \vec{\alpha} \in \mathbb{N}_0^n,$$
$$\sum_{l \in \mathbb{N}_0} \tilde{\eta}_l(\vec{k}) = 1.$$

A Function Spaces

For example, let $\tilde{\eta} \in \mathcal{S}(\mathbb{R}^n)$ with

$$\tilde{\eta}(\vec{k}) = 1 \text{ if } |\vec{k}|_\infty \leq 1 \text{ and } \tilde{\eta}(\vec{k}) = 0 \text{ if } |\vec{k}|_\infty \geq 2 \tag{A.1}$$

and let further

$$\tilde{\eta}_0(\vec{k}) := \tilde{\eta}(\vec{k}) \quad \text{and} \quad \tilde{\eta}_l(\vec{k}) := \tilde{\eta}(2^{-l}\vec{k}) - \tilde{\eta}(2^{-(l-1)}\vec{k}), l \in \mathbb{N}.$$

Then, the family $\{\tilde{\eta}_l\}_{l \in \mathbb{N}_0}$ forms an inhomogeneous smooth dyadic partition of unity.

Definition A.4. *Let $1 \leq p < \infty$, $1 \leq q < \infty$, $r \in \mathbb{R}$. Then, the (inhomogeneous) Besov space is defined by*

$$B^r_{p,q}(\mathbb{R}^n) := \left\{ f \in \mathcal{S}'(\mathbb{R}^n) \,:\, \|f\|_{B^r_{p,q}} := \left(\sum_{l \in \mathbb{N}_0} 2^{lrq} \left\| \mathcal{F}^{-1}[\tilde{\eta}_l \hat{f}] \right\|_{\mathcal{L}^p}^q \right)^{\frac{1}{q}} < \infty \right\}.$$

Definition A.5. *Let $1 \leq p < \infty$, $1 \leq q < \infty$, $r \in \mathbb{R}$. Then, the (inhomogeneous) Triebel-Lizorkin space is defined by*

$$F^r_{p,q}(\mathbb{R}^n) := \left\{ f \in \mathcal{S}'(\mathbb{R}^n) \,:\, \|f\|_{B^r_{p,q}} := \left\| \left(\sum_{l \in \mathbb{N}_0} 2^{lrq} \left| \mathcal{F}^{-1}[\tilde{\eta}_l \hat{f}] \right|^q \right)^{\frac{1}{q}} \right\|_{\mathcal{L}^p} < \infty \right\}.$$

Let us remark that these spaces are independent of the dyadic partition of unity $\{\tilde{\eta}_l\}_{l \in \mathbb{N}_0}$ and that for $p = q = 2$ we particularly have $\mathcal{H}^r(\mathbb{R}^n) = B^r_{2,2}(\mathbb{R}^n) = F^r_{2,2}(\mathbb{R}^n)$ up to norm equivalence. Note that one may extend the parameter sets for the Besov spaces to $0 < p, q \leq \infty$ and for the Triebel-Lizorkin spaces to $0 < p < \infty$ and $0 < q \leq \infty$. Note, further, that the so-called homogeneous Besov spaces and homogeneous Triebel-Lizorkin spaces are defined in a similar way by a homogeneous smooth dyadic partition of unity instead of an inhomogeneous one. For example, let $\tilde{\eta} \in \mathcal{S}(\mathbb{R}^n)$ given by (A.1) and let the family $\{\tilde{\eta}_l\}_{l \in \mathbb{Z}}$ be given by

$$\tilde{\eta}_l(\vec{k}) := \tilde{\eta}(2^{-l}\vec{k}) - \tilde{\eta}(2^{-(l-1)}\vec{k}), l \in \mathbb{Z}.$$

For a further reading see e.g. [117, 178]. Furthermore, with regard to Besov and Triebel-Lizorkin spaces of dominating mixed smoothness see [162, 179].

B Wavelets

In the following we recall a few details in the framework of univariate and multivariate wavelets and review in particular the univariate Meyer wavelet family. For a further reading see e.g. [32, 34, 37, 42, 87, 130, 139, 178].

B.1 Multiresolution analysis and multivariate wavelets

Let us recall the definition of a multiresolution analysis on $\mathbb{R}$. We consider an infinite sequence

$$\cdots \subset V_{-2} \subset V_{-1} \subset V_0 \subset V_1 \subset V_2 \subset \cdots \qquad (B.1)$$

of nested spaces V_l with $\bigcap_{l \in \mathbb{Z}} V_l = 0$ and $\operatorname{clos}_{\mathcal{L}^2} \bigcup_{l \in \mathbb{Z}} V_l = \mathcal{L}^2(\mathbb{R})$. Then there holds $f(x) \in V_l \Leftrightarrow f(2x) \in V_{l+1}$ and $f(x) \in V_0 \Leftrightarrow f(x-j) \in V_0$, where $j \in \mathbb{Z}$. Furthermore, there is a so-called scaling function (or father wavelet) $\varphi \in V_0$, such that $\{\varphi(x-j) : j \in \mathbb{Z}\}$ forms an orthonormal basis for V_0. Then,

$$\{\varphi_{l,j}(x) = 2^{\frac{l}{2}} \varphi(2^l x - j) : j \in \mathbb{Z}\}$$

forms an orthonormal basis of V_l and we can represent any $f(x) \in V_l$ as $f(x) = \sum_{j=-\infty}^{\infty} \langle \varphi_{l,j}^*, f \rangle \varphi_{l,j}(x)$. With the definition

$$V_l \oplus W_l = V_{l+1}, \quad W_l \perp V_l, \qquad (B.2)$$

we obtain an associated sequence of detail spaces W_l with associated mother wavelet $\psi \in W_0$, such that $\{\psi(x-j) : j \in \mathbb{Z}\}$ forms an orthonormal basis for W_0. Thus,

$$\{\psi_{l,j}(x) = 2^{\frac{l}{2}} \psi(2^l x - j) : j \in \mathbb{Z}\}$$

gives an orthonormal basis for W_l and $\{\psi_{l,j} : l, j \in \mathbb{Z}\}$ is an orthonormal basis of $\mathcal{L}^2(\mathbb{R})$. Then, in particular, the orthogonal direct sum decomposition

$$\mathcal{L}^2(\mathbb{R}) = \bigoplus_{l \in \mathbb{Z}} W_l \qquad (B.3)$$

holds and we can represent any $f(x)$ in $\mathcal{L}^2(\mathbb{R})$ as

$$f(x) = \sum_{l=-\infty}^{\infty} \sum_{j=-\infty}^{\infty} \langle \psi_{l,j}, f \rangle \psi_{l,j}(x). \qquad (B.4)$$

Furthermore, for a chosen coarsest level $L_0 \in \mathbb{Z}$ we can reformulate (B.3) and (B.4) to the orthogonal direct sum decomposition

$$\mathcal{L}^2(\mathbb{R}) = V_{L_0} \oplus \bigoplus_{l \in \mathbb{Z}, l \geq L_0} W_l \qquad (B.5)$$

B Wavelets

and the representation

$$f(x) = \sum_{j=-\infty}^{\infty} \langle \varphi_{L_0,j}, f \rangle \varphi_{L_0,j}(x) + \sum_{l=L_0}^{\infty} \sum_{j=-\infty}^{\infty} \langle \psi_{l,j}, f \rangle \psi_{l,j}(x), \tag{B.6}$$

respectively.

Additionally, we can scale our spaces and decompositions by a parameter $c > 0$, $c \in \mathbb{R}$. For example, we can set $V_l^c = \text{clos}_{\mathcal{L}^2}(\text{span}\{\varphi_{c,l,j}(x) = c^{\frac{1}{2}} 2^{\frac{l}{2}} \varphi(c2^l x - j) : j \in \mathbb{Z}\})$. For $c = 2^k$, $k \in \mathbb{Z}$, the obvious identity $V_l^c = V_{l+k}^1$ holds. Then, we obtain the scaled decomposition

$$V_L^c = V_{L_0}^c \oplus \bigoplus_{l=L_0}^{L} W_l^c$$

with the scaled detail spaces $W_l^c = \text{clos}_{\mathcal{L}^2}(\text{span}\{\psi_{c,l,j}(x) = c^{\frac{1}{2}} 2^{\frac{l}{2}} \psi(c2^l x - j) : j \in \mathbb{Z}\})$. For $c = 2^k$, $k \in \mathbb{Z}$, the identity $W_l^c = W_{l+k}^1$ holds. With the choice $c = 2^{-L_0}$ we can get discard the parameter L_0 and write our orthogonal direct sum decomposition (B.5) as $V_0^c \oplus \bigoplus_{l=0}^{\infty} W_l^c$. To simplify notation we introduce

$$\phi_{l,j} := \begin{cases} \varphi_{l,j}^c & l = 0, \\ \psi_{l-1,j}^c & l \geq 1 \end{cases} \tag{B.7}$$

for $c = 2^{-L_0}$. In this way, we skip the scaling index c and obtain a unique notation for both the father wavelets on the coarsest scale and the mother wavelets of the detail spaces. In the following, bear in mind that the function $\phi_{l,j}$ with $l = 0$ denotes a father wavelet, i.e. a scaling function only, whereas for $l \geq 1$ it denotes a true wavelet on scale $l - 1$. Then, the family $\{\phi_{l,j}\}_{l \in \mathbb{N}_0, j \in \mathbb{Z}}$ is an orthonormal basis of $\mathcal{L}^2(\mathbb{R}) = V_0 \oplus \bigoplus_{l \in \mathbb{N}_0} W_l$.

In order to obtain a multivariate basis for $\mathcal{L}^2(\mathbb{R}^n)$ starting from a univariate wavelet basis set $\{\phi_{l,j}\}_{l \in \mathbb{N}_0, j \in \mathbb{Z}}$ as in (B.7), there are at least two ways based on tensor product construction. First, there is a trivial anisotropic construction, where simply the tensor product functions

$$\mathcal{B}_{\text{aniso}} := \left\{ \bigotimes_{\mu=1}^{n} \phi_{l_\mu, j_\mu} \right\}_{\vec{l} \in \mathbb{N}_0^n, \vec{k} \in \mathbb{Z}^n}$$

are taken. Second, there is an isotropic way to construct a multivariate basis for $\mathcal{L}^2(\mathbb{R}^n)$. Here, to simplify notation, we introduce the univariate functions

$$\phi_{l,j}^{[z]} := \begin{cases} \varphi_{l,j}^c & z = 0, \\ \psi_{l,j}^c & z = 1 \end{cases}$$

for $z \in \{0, 1\}$, $l \in \mathbb{N}_0$, $j \in \mathbb{Z}$. Now, we set $\varphi_{\vec{j}}(\vec{x}) := \bigotimes_{\mu=1}^{N} \phi_{0, j_\mu}^{[0]}(\vec{x})$ and $\psi_{l,\vec{j}}^{[\vec{z}]}(\vec{x}) := \bigotimes_{\mu=1}^{N} \phi_{l, j_\mu}^{[z_\mu]}(\vec{x})$ for $\vec{z} \in \mathcal{Z} := \{0, 1\}^n \setminus \vec{0}$. Then, the multivariate functions

$$\mathcal{B}_{\text{iso}} := \{\varphi_{\vec{j}}\}_{\vec{j} \in \mathbb{Z}^n} \cup \bigcup_{\vec{z} \in \mathcal{Z}} \{\psi_{l,\vec{j}}^{[\vec{z}]}\}_{l \in \mathbb{N}_0, \vec{j} \in \mathbb{Z}^D}$$

build an isotropic multivariate basis for $\mathcal{L}^2(\mathbb{R}^n)$, which can be seen as the tensor product of multiresolution analyses. That is, starting form the univariate case according to (B.1) and (B.2) we consider the sequence of spaces

$$\bigotimes_{\mu=1}^{n} V_0 \subset \bigotimes_{\mu=1}^{n} V_1 \subset \bigotimes_{\mu=1}^{n} V_2 \subset \cdots$$

and the subspace splitting

$$\bigotimes_{\mu=1}^{n} V_{l+1} = \bigotimes_{\mu=1}^{n}(V_l \oplus W_l) = \bigotimes_{\mu=1}^{n} V_l \oplus \bigoplus_{\vec{z} \in \mathcal{Z}} W_l^{[\vec{z}]},$$

respectively, where $W_l^{[\vec{z}]}$ denotes the tensor product space

$$W_l^{[\vec{z}]} := \bigotimes_{\mu=1}^{n} U_l^{[z]}, \quad U_l^{[z]} := \begin{cases} V_l, & z = 0 \\ W_l & z = 1. \end{cases}$$

Then, we have the orthogonal direct sum decomposition

$$\mathcal{L}^2(\mathbb{R}^n) = \bigotimes_{\mu=1}^{n} V_0 \oplus \bigoplus_{l \in \mathbb{N}_0} \bigoplus_{\vec{z} \in \mathcal{Z}} W_l^{[\vec{z}]}$$

and for any $f \in \mathcal{L}^2(\mathbb{R}^n)$ the representation

$$f(\vec{x}) = \sum_{\vec{j} \in \mathbb{Z}^n} \langle \varphi_j, f \rangle \varphi_j(x) + \sum_{l \in \mathbb{N}_0} \sum_{\vec{z} \in \mathcal{Z}} \sum_{\vec{j} \in \mathbb{Z}^n} \langle \psi_{l,j}^{[\vec{z}]}, f \rangle \psi_{l,j}^{[\vec{z}]}(x),$$

according to (B.5) and (B.6).

B.2 Meyer wavelet family

Definition B.1. *According to [139], we set as father and mother wavelet in Fourier space*

$$\hat{\varphi}(k) = \frac{1}{\sqrt{2\pi}} \begin{cases} 1 & \text{for } |k| \leq \frac{2}{3}\pi, \\ \cos(\frac{\pi}{2}\rho(\frac{3}{2\pi}|k|-1)) & \text{for } \frac{2\pi}{3} < |k| \leq \frac{4\pi}{3}, \\ 0 & \text{otherwise,} \end{cases} \tag{B.8}$$

$$\hat{\psi}(k) = \frac{1}{\sqrt{2\pi}} e^{-i\frac{k}{2}} \begin{cases} \sin(\frac{\pi}{2}\rho(\frac{3}{2\pi}|k|-1)) & \text{for } \frac{2}{3}\pi \leq |k| \leq \frac{4}{3}\pi, \\ \cos(\frac{\pi}{2}\rho(\frac{3}{4\pi}|k|-1)) & \text{for } \frac{4\pi}{3} < |k| \leq \frac{8\pi}{3}, \\ 0 & \text{otherwise,} \end{cases} \tag{B.9}$$

where $\rho: \mathbb{R} \to \mathbb{R} \in C^r$ is a parameter function still to be fixed, with properties $\rho(x) = 0$ for $x \leq 0$, $\rho(x) = 1$ for $x > 1$ and $\rho(x) + \rho(1-x) = 1$.

For $c > 0$ by dilation and translation we obtain

$$\mathcal{F}[\varphi_{c,l,j}](k) = \hat{\varphi}_{c,l,j}(k) = c^{-\frac{1}{2}} 2^{-\frac{l}{2}} e^{-ic^{-1} 2^{-l} jk} \hat{\varphi}(c^{-1} 2^{-l} k),$$
$$\mathcal{F}[\psi_{c,l,j}](k) = \hat{\psi}_{c,l,j}(k) = c^{-\frac{1}{2}} 2^{-\frac{l}{2}} e^{-ic^{-1} 2^{-l} jk} \hat{\psi}(c^{-1} 2^{-l} k),$$

B Wavelets

where the $\hat{\varphi}_{c,l,j}$ and $\hat{\psi}_{c,l,j}$ denote the dilates and translates of (B.8) and (B.9), respectively.

This wavelet family can be derived from a partition of unity $\sum_{l\in\mathbb{N}_0} \eta_{c,l}(k) = 1, \forall k \in \mathbb{R}$ in Fourier space, where

$$\eta_{c,l}(k) = \begin{cases} c 2\pi \hat{\varphi}^*_{c,0,0}(k)\hat{\varphi}_{c,0,0}(k) & \text{for } l = 0, \\ c 2^l \pi \hat{\psi}^*_{c,l-1,0}(k)\hat{\psi}_{c,l-1,0}(k) & \text{for } l > 0. \end{cases} \quad (B.10)$$

See [6, 139] for details. The function ρ basically describes the decay from one to zero of one partition function $\eta_{c,l}$ in the overlap with its neighbor. The smoothness of the $\eta_{c,l}$ is thus directly determined by the smoothness of ρ. The mother wavelets $\hat{\psi}_{c,l,j}$ and the father wavelets $\hat{\varphi}_{c,l,j}$ in Fourier space inherit the smoothness of the $\eta_{c,l}$ via relation (B.10).

There are various choices for ρ with different smoothness properties in literature; see [37, 139, 183]. Examples are the Shannon wavelet and the raised cosine wavelet, i.e. (B.9) with

$$\rho(x) = \rho^0(x) := \begin{cases} 0 & x \leq \frac{1}{2}, \\ 1 & \text{otherwise} \end{cases} \quad \text{and} \quad \rho(x) = \rho^1(x) \begin{cases} 0 & x \leq 0, \\ x & 0 \leq x \leq 1, \\ 1 & \text{otherwise} \end{cases} \quad (B.11)$$

and

$$\rho(x) = \rho^\infty(x) := \begin{cases} 0 & x \leq 0, \\ \frac{\tilde{\rho}(x)}{\tilde{\rho}(1-x)+\tilde{\rho}(x)} & 0 < x \leq 1, \\ 1 & \text{otherwise,} \end{cases} \quad \text{where} \quad \tilde{\rho}(x) = \begin{cases} 0 & x \leq 0, \\ e^{-\frac{1}{x^2}} & \text{otherwise,} \end{cases}$$

(B.12)

respectively. Other types of Meyer wavelets with different smoothness properties can be found in [6, 87, 103, 190]. The resulting mother wavelet functions in real space and Fourier space are given in Figure B.1. Note the two symmetric areas of support and the associated two bands with non-zero values of the wavelets in Fourier space, which resemble the line of construction due to Wilson, Malvar, Coifman and Meyer to circumvent the Balian-Low theorem; see e.g. [138].

Let us remark here, that the Balian-Low theorem basically states that the family of functions $g_{m,n}(x) = e^{2\pi i m x} g(x-n)$, $m, n \in \mathbb{Z}$, which are related to the windowed Fourier transform, cannot be an orthonormal basis of $\mathcal{L}^2(\mathbb{R})$ if the two integrals $\int_\mathbb{R} x^2 |g(x)|^2 dx$ and $\int_\mathbb{R} k^2 |\hat{g}(k)|^2 dk$ are both finite. Thus, there exists no orthonormal family for a Gaussian window function $g(x) = \pi^{-1/4} e^{-x^2/2}$ which is both sufficiently regular and well localized [87].

In real space, these wavelets are C^∞-functions with global support; in Fourier space, they are piecewise continuous, piecewise continuous differentiable and C^∞, respectively, and have compact support. Furthermore, they possess infinitely many vanishing moments. Finally, their envelope in real space decays with $|x| \to \infty$ as $|x|^{-1}$ for ρ^0, as $|x|^{-2}$ for ρ^1 and faster than any polynomial (sub-exponentially) for ρ^∞, respectively. To our knowledge, only for the Meyer wavelets with (B.11) there are analytical formulae available in both real and Fourier space available. Certain integrals in a Galerkin discretization of (3.16) can then be given analytically. For the other types of Meyer wavelets analytical formulae only exist in Fourier space and therefore, numerical integration is necessary in a Galerkin discretization of (3.16).

B.2 Meyer wavelet family

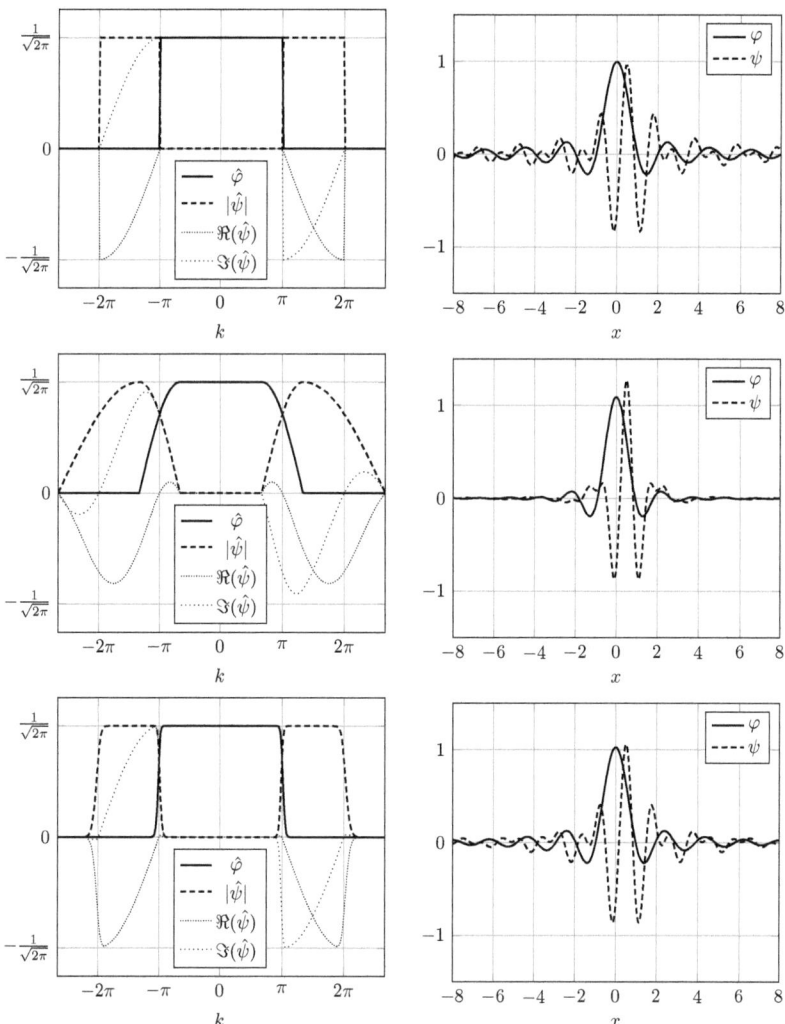

Figure B.1: Top: (B.9) with ρ^0 from (B.11) in Fourier space (left) and real space (right). Middle: (B.9) with ρ^1 from (B.11) in Fourier space (left) and real space (right). Bottom: (B.9) with ρ^∞ from (B.12) in Fourier space (left) and real space (right).

B Wavelets

C One- and Two-Particle Operator Integrals

In the following we recall analytic formulae with respect to the one- and two-particle operator integrals according to Section 4.1.1 involving isotropic Gaussians (4.17); see e.g. [136, 170]. Furthermore, we recall analogous analytic formulae involving modulated isotropic Gaussians (4.38); see e.g. [3, 134, 171].

C.1 Integrals involving Gaussians

First, let us note that there holds the so-called product theorem

$$e^{-a|\mathbf{x}-\mathbf{R}_a|_2^2} e^{-b|\mathbf{x}-\mathbf{R}_b|_2^2} = e^{-P} e^{-p|\mathbf{x}-\mathbf{R}_p|_2^2},$$

where $p = a + b$, $\mathbf{R}_p = \frac{a}{a+b}\mathbf{R}_a + \frac{b}{a+b}\mathbf{R}_b$ and $P = \frac{ab}{a+b}|\mathbf{R}_a - \mathbf{R}_b|_2^2$.

Now, let $a > 0$, $b > 0$ and $\mathbf{R}_a, \mathbf{R}_b \in \mathbb{R}^3$, then the overlap integrals can be written in the form

$$\int_{\mathbb{R}^3} e^{-a(\mathbf{x}-\mathbf{R}_a)^2} e^{-b(\mathbf{x}-\vec{R}_b)^2} d\mathbf{x} = \left(\frac{\pi}{a+b}\right)^{\frac{3}{2}} e^{-\frac{ab}{a+b}|\mathbf{R}_a-\mathbf{R}_b|_2^2}. \tag{C.1}$$

The one-particle operator integrals with respect to the kinetic energy operator $-\frac{1}{2}\Delta_{\mathbf{x}}$ can be written as

$$\frac{1}{2}\int_{\mathbb{R}^3} \nabla_{\mathbf{x}}\left(e^{-a|\mathbf{x}-\mathbf{R}_a|_2^2}\right) \nabla_{\mathbf{x}}\left(e^{-b|\mathbf{x}-\mathbf{R}_b|_2^2}\right) d\mathbf{x} =$$

$$\frac{ab}{a+b}\left(3 - \frac{2ab}{a+b}|\mathbf{R}_a - \mathbf{R}_b|_2^2\right)\left(\frac{\pi}{a+b}\right)^{\frac{3}{2}} e^{-\frac{ab}{a+b}|\mathbf{R}_a-\mathbf{R}_b|_2^2},$$

and the one-particle operator integrals with respect to the electron-nuclei Coulomb potential $-Z/|\mathbf{x}-\mathbf{R}|_2$, $\mathbf{R} \in \mathbb{R}^3$ can be written in the form

$$-\int_{\mathbb{R}^3} e^{-a|\mathbf{x}-\mathbf{R}_a|_2^2} \frac{Z}{|\mathbf{x}-\mathbf{R}|} e^{-b|\mathbf{x}-\mathbf{R}_b|_2^2} d\mathbf{x} = -\frac{2\pi Z}{a+b} F_0\left[(a+b)\left|\mathbf{R}-\mathbf{R}_p\right|_2^2\right] e^{-P}. \tag{C.2}$$

Here, $F_m(x)$ denotes the incomplete Gamma function of integer order m, where for $m = 0$ we particularly have the identity

$$F_0(x) = \frac{1}{2}\left(\frac{\pi}{t}\right)^{\frac{1}{2}} \mathrm{erf}(t^{\frac{1}{2}}).$$

The two-particle operator integral can be written in the form

$$\int_{\mathbb{R}^{3\times 3}} e^{-a|\mathbf{x}-\mathbf{R}_a|_2^2} e^{-c|\mathbf{y}-\mathbf{R}_c|_2^2} \frac{1}{|\mathbf{x}-\mathbf{y}|} e^{-b|\mathbf{x}-\mathbf{R}_b|_2^2} e^{-d|\mathbf{y}-\mathbf{R}_d|_2^2} d\mathbf{x}\, d\mathbf{y} =$$

$$\frac{2\pi^{\frac{5}{2}}}{pq\sqrt{p+q}} F_0\left[\frac{pq}{p+q}|\mathbf{R}_p - \mathbf{R}_q|_2^2\right] e^{-P} e^{-Q},$$

C One- and Two-Particle Operator Integrals

where again $p = a+b$, $\mathbf{R}_p = \frac{a}{a+b}\mathbf{R}_a + \frac{b}{a+b}\mathbf{R}_b$, $P = \frac{ab}{a+b}|\mathbf{R}_a - \mathbf{R}_b|_2^2$ and, in addition, $q = c+d$, $\mathbf{R}_q = \frac{c}{c+d}\mathbf{R}_d + \frac{d}{c+d}\mathbf{R}_d$ and $P = \frac{cd}{c+d}|\mathbf{R}_c - \mathbf{R}_d|_2^2$.

Note that in the more general case of Gauss-Hermite functions one can apply the so-called McMurchie-Davidson recursive formulae, which then involve the incomplete Gamma functions of higher order m; see [136].

C.2 Integrals involving modulated Gaussians

A modified Gaussian can be expressed in the form of an isotropic Gaussian with a complex center, i.e. there holds the easily shown relation

$$e^{i\mathbf{k}_a^T \mathbf{x}} e^{-a|\mathbf{x}-\mathbf{R}_a|_2^2} = e^{i\mathbf{k}_a^T(\mathbf{R}_a - \frac{1}{4a}\mathbf{k}_a)} e^{-a|\mathbf{x}-(\mathbf{R}_a + i\frac{1}{2a}\mathbf{k}_a)|_2^2}.$$

With its help one can deduce analytic formulae with regard to the necessary one- and two-operator integrals in an analogous way to the case of real isotropic Gaussians [3, 171]. For example, analogously to (C.1) one obtains the identity

$$\int_{\mathbb{R}^3} e^{i\mathbf{k}_a^T \mathbf{x}} e^{-a(\mathbf{x}-\mathbf{R}_a)^2} e^{i\mathbf{k}_b^T \mathbf{x}} e^{-b(\mathbf{x}-\vec{R}_b)^2} d\mathbf{x} =$$

$$e^{i\mathbf{k}_a^T(\mathbf{R}_a - \frac{1}{4a}\mathbf{k}_a)} e^{i\mathbf{k}_b^T(\mathbf{R}_b - \frac{1}{4b}\mathbf{k}_b)} \left(\frac{\pi}{a+b}\right)^{\frac{3}{2}} e^{-\frac{ab}{a+b}|\mathbf{R}_a + i\frac{1}{2a}\mathbf{k}_a - (\mathbf{R}_b + i\frac{1}{2b}\mathbf{k}_b)|_2^2}$$

and analogously to (C.2) the relation

$$-\int_{\mathbb{R}^3} e^{i\mathbf{k}_a^T \mathbf{x}} e^{-a(\mathbf{x}-\mathbf{R}_a)^2} \frac{Z}{|\mathbf{x}-\mathbf{R}|_2} e^{i\mathbf{k}_b^T \mathbf{x}} e^{-b(\mathbf{x}-\vec{R}_b)^2} d\mathbf{x} = -e^{i\mathbf{k}_a^T(\mathbf{R}_a - \frac{1}{4a}\mathbf{k}_a)} e^{i\mathbf{k}_b^T(\mathbf{R}_b - \frac{1}{4b}\mathbf{k}_b)}$$

$$\cdot \frac{2\pi Z}{a+b} F_0 \left[(a+b)\left|\mathbf{R} - (\mathbf{R}_a + i\frac{1}{2a}\mathbf{k}_a - (\mathbf{R}_b + i\frac{1}{2b}\mathbf{k}_b))\right|_2^2\right] e^{-P}.$$

Here, one has to evaluate the complex error function in the case of the one- and two-particle Coulomb integrals. In this thesis we use the identity

$$\operatorname{erf}(z) = 1 - e^{-z^2} w(iz), \quad z \in \mathbb{C}$$

together with an algorithm to compute the so-called Faddeeva function w. This algorithm is given in [153] and is based on an algorithm given in [61].

In the case of the application of Gauss-Hermite functions instead of isotropic Gaussians, one can employ recursion formulae similar to that of McMurchie-Davidson [171]. Then, the complex incomplete Gamma function has to be evaluated. For a further reading on an efficient computation of the complex incomplete Gamma function; see e.g. [99, 132].

Bibliography

[1] R. ADAMS, *Sobolev Spaces*, Academic Press, 1975.

[2] S. AGMON, *Lectures on Exponential Decay of Solutions of Second-order Elliptic Equations: Bounds on Eigenfunctions of N-body Schrödinger Operators*, vol. 29 of Mathematical Notes, Princeton University Press, 1982.

[3] D. ALLISON, N. HANDY, AND S. BOYS, *A new basis set for molecular wavefunctions*, Molecular Physics, 26 (1973), pp. 715–723.

[4] E. ANDERSON, Z. BAI, C. BISCHOF, L. BLACKFORD, J. DEMMEL, J. DONGARRA, J. D. CROZ, S. HAMMARLING, A. GREENBAUM, A. MCKENNEY, AND D. SORENSEN, *LAPACK Users' guide*, SIAM, 3 ed., 1999.

[5] T. ARIAS, *Multiresolution analysis of electronic structure: Semicardinal and wavelet bases*, Reviews of Modern Physics, 71 (1999), pp. 267–311.

[6] P. AUSCHER, G. WEISS, AND M. WICKERHAUSER, *Local sine and cosine bases of Coifman and Meyer and the construction of smooth wavelets*, in Wavelets: A Tutorial in Theory and Applications, C. Chui, ed., vol. 2 of Wavelet Analysis and Its Applications, Academic Press, 1992, pp. 237–256.

[7] Z. BAI, J. DEMMEL, J. DONGARRA, A. RUHE, AND H. VAN DER VORST, eds., *Templates for the Solution of Algebraic Eigenvalue Problems: A Practical Guide*, SIAM, 2000.

[8] S. BALAY, K. BUSCHELMAN, V. EIJKHOUT, W. GROPP, D. KAUSHIK, M. KNEPLEY, L. MCINNES, B. SMITH, AND H. ZHANG, *PETSc users manual*, Tech. Rep. ANL-95/11 - Revision 2.1.5, Argonne National Laboratory, 2004.

[9] S. BALAY, K. BUSCHELMAN, W. GROPP, D. KAUSHIK, M. KNEPLEY, L. MCINNES, B. SMITH, AND H. ZHANG, *PETSc Web page*, 2001. http://www.mcs.anl.gov/petsc.

[10] S. BALAY, W. GROPP, L. MCINNES, AND B. SMITH, *Efficient management of parallelism in object oriented numerical software libraries*, in Modern Software Tools in Scientific Computing, E. Arge, A. M. Bruaset, and H. P. Langtangen, eds., Birkhäuser Press, 1997, pp. 163–202.

[11] R. BARROIS, S. BEKAVAC, AND H. KLEINDIENST, *Nonrelativistic energies for the lowest $^4P^0$ degrees states of the Li isoelectronic series*, Chemical Physics Letters, 268 (1997), pp. 531–534.

[12] R. BELLMANN, *Adaptive Control Processes: A Guided Tour*, Princeton University Press, 1961.

Bibliography

[13] G. BEYLKIN AND M. MOHLENKAMP, *Numerical operator calculus in higher dimensions*, Proceedings of the National Academy of Sciences of the United States of America, 99 (2002), pp. 10246–10251.

[14] ———, *Algorithms for numerical analysis in high dimensions*, SIAM Journal on Scientific Computing, 26 (2005), pp. 2133–2159.

[15] G. BEYLKIN, M. MOHLENKAMP, AND F. PÉREZ, *Approximating a wavefunction as an unconstrained sum of Slater determinants*, Journal of Mathematical Physics, 49 (2008), p. 032107.

[16] A. BITTNER, F. HEBER, AND J. HAMAEKERS, *Biomolecules as soft matter surfaces*, Surface Science, 603 (2009), pp. 1922–1925.

[17] S. BOYS, *The integral formulae for the variational solution of the molecular many-electron wave equation in terms of gaussian functions with direct electronic correlation*, Proceedings of the Royal Society of London Series A: Mathematical and Physical Sciences, 258 (1960), pp. 402–411.

[18] S. BOYS AND N. HANDY, *Determination of energies and wavefunctions with full electronic correlation*, Proceedings of the Royal Society of London Series A: Mathematical and Physical Sciences, 310 (1969), pp. 43–61.

[19] D. BRAESS, *Asymptotics for the approximation of wave functions by exponential sums*, Journal of Approximation Theory, 83 (1995), pp. 93–103.

[20] D. BRAESS AND W. HACKBUSCH, *Approximation of $1/x$ by exponential sums in $[1, \infty)$*, IMA Journal of Numerical Analysis, 25 (2005), pp. 685–697.

[21] M. BROWN, J. TRAIL, P. RIOS, AND R. NEEDS, *Energies of the first row atoms from quantum monte carlo*, Journal of Chemical Physics, 126 (2007), p. 224110.

[22] H.-Q. BUI AND R. LAUGESEN, *Sobolev spaces and approximation by affine spanning systems*, Mathematische Annalen, 341 (2008), pp. 347–389.

[23] H. BUNGARTZ AND M. GRIEBEL, *A note on the complexity of solving Poisson's equation for spaces of bounded mixed derivatives*, J. Complexity, 15 (1999), pp. 167–199.

[24] ———, *Sparse grids*, Acta Numerica, 13 (2004), pp. 1–123.

[25] E. CANCÉS, M. DEFRANCESCHI, W. KUTZELNIGG, C. LE BRIS, AND Y. MADAY, *Computational quantum chemistry: A primer*, in Special Volume: Computational Chemistry., P. Ciarlet and C. Le Bris, eds., vol. X of Handbook of Numerical Analysis, Elsevier Science B.V., 2003, ch. 1, pp. 1–270.

[26] M. CASULA, C. ATTACCALITE, AND S. SORELLA, *Correlated geminal wave function for molecules: An efficient resonating valence bond approach*, Journal of Chemical Physics, 121 (2004), pp. 7110–7126.

[27] W. CENCEK, J. KOMASA, AND J. RYCHLEWSKI, *Benchmark calculations for 2-electron systems using explicitly correlated Gaussian functions*, Chemical Physics Letters, 246 (1995), pp. 417–420.

[28] W. CENCEK AND J. RYCHLEWSKI, *Benchmark calculations for He_2^+ and LiH molecules using explicitly correlated gaussian functions*, Chemical Physics Letters, 320 (2000), pp. 549–552.

[29] S. CHAKRAVORTY, S. GWALTNEY, E. DAVIDSON, F. PARPIA, AND C. FISCHER, *Ground-state correlation energies for atomic ions with 3 to 18 electrons*, Physical Review A, 47 (1993), pp. 3649–3670.

[30] S. CHINNAMSETTY, M., ESPIG, B. KHOROMSKIJ, W. HACKBUSCH, AND H.-J. FLAD, *Tensor product approximation with optimal rank in quantum chemistry*, Journal of Chemical Physics, 127 (2007), p. 084110.

[31] K. CHO, T. ARIAS, J. JOANNOPOULOS, AND P. LAM, *Wavelets in electronic-structure calculations*, Physical Review Letters, 71 (1993), pp. 1808–1811.

[32] C. CHUI, *An Introduction to Wavelets*, vol. 1 of Wavelet Analysis And Its Applications, Academic Press, 1992.

[33] C. CHUI, J. STOCKLER, AND J. WARD, *Analytic wavelets generated by radial functions*, Advances in Computational Mathematics, 5 (1996), pp. 95–123.

[34] A. COHEN, W. DAHMEN, AND R. DEVORE, *Adaptive wavelet techniques in numerical simulation*, in Encyclopedia of Computational Mechanics, R. De Borste, T. Hughes, and E. Stein, eds., vol. 1, John Wiley & Sons, 2004, pp. 157–197.

[35] E. CONDON, *The theory of complex spectra*, Physical Review, 36 (1930), pp. 1121–1133.

[36] T. CRAWFORD, C. SHERRILL, E. VALEEV, J. FERMANN, R. KING, M. LEININGER, S. BROWN, C. JANSSEN, E. SEIDL, J. KENNY, AND W. ALLEN, *PSI3: An open-source ab initio electronic structure package*, Journal of Computational Chemistry, 28 (2007), pp. 1610–1616.

[37] I. DAUBECHIES, *Ten lectures on wavelets*, vol. 61 of CBMS-NSF Regional Conference Series in Applied Mathematics, SIAM, 1992.

[38] E. DAVIDSON, S. HAGSTROM, S. CHAKRAVORTY, V. UMAR, AND C. FISCHER, *Ground-state correlation energies for 2-electron to 10-electron atomic ions*, Physical Review A, 44 (1991), pp. 7071–7083.

[39] C. DE BOOR, R. DEVORE, AND A. RON, *Approximation orders of FSI spaces in $L^2(\mathbb{R}^D)$*, Constructive Approximation, 14 (1998), pp. 631–652.

[40] S. DEKEL AND D. LEVIATAN, *Wavelet decompositions of nonrefinable shift invariant spaces*, Applied and Computational Harmonic Analysis, 12 (2002), pp. 230–258.

Bibliography

[41] R. DEVORE, *Nonlinear approximation*, Acta Numerica, 7 (1998), pp. 51–150.

[42] R. DEVORE AND B. LUCIER, *Wavelets*, Acta Numerica, 1 (1992), pp. 1–56.

[43] J. DICK, I. SLOAN, X. WANG, AND H. WOŹNIAKOWSKI, *Liberating the weights*, Jounal of Complexity, 20 (2004), pp. 593–623.

[44] F. DIJKSTRA AND J. V. LENTHE, *On the rapid evaluation of cofactors in the calculation of nonorthogonal matrix elements*, International Journal of Quantum Chemistry, 67 (1998), pp. 77–83.

[45] P. DINTELMANN, *Classes of Fourier multipliers and Besov-Nikolskij spaces*, Mathematische Nachrichten, 173 (1995), pp. 115–130.

[46] D. DONOHO, *Tight frames of k-plane ridgelets and the problem of representing objects that are smooth away from d-dimensional singularities in $\mathbb{R}^n$*, Proceedings of the National Academy of Sciences of the United States of America, 96 (1999), pp. 1828–1833.

[47] T. DUNNING, *Gaussian-basis sets for use in correlated molecular calculations. 1. The atoms boron through neon and hydrogen*, Journal of Chemical Physics, 90 (1989), pp. 1007–1023.

[48] B. EFRON AND C. STEIN, *The jackknife estimate of variance*, The Annals of Statistics, 9 (1981), pp. 586–596.

[49] C. FILIPPI AND C. UMRIGAR, *Multiconfiguration wave functions for quantum Monte Carlo calculations of first-row diatomic molecules*, Journal of Chemical Physics, 105 (1996), pp. 213–226.

[50] H. FLAD, W. HACKBUSCH, D. KOLB, AND R. SCHNEIDER, *Wavelet approximation of correlated wave functions. I. Basics*, Journal of Chemical Physics, 116 (2002), pp. 9641–9657.

[51] H. FLAD, W. HACKBUSCH, AND R. SCHNEIDER, *Best N-term approximation in electronic structure calculations - I. One-electron reduced density matrix*, Mathematical Modelling and Numerical Analysis, 40 (2006), pp. 49–61.

[52] ———, *Best N-term approximation in electronic structure calculations. II. Jastrow factors*, Mathematical Modelling and Numerical Analysis, 41 (2007), pp. 261–279.

[53] W. FOULKES, L. MITAS, R. NEEDS, AND G. RAJAGOPAL, *Quantum Monte Carlo simulations of solids*, Reviews of Modern Physics, 73 (2001), pp. 33–83.

[54] S. FOURNAIS, M. HOFFMANN-OSTENHOF, T. HOFFMANN-OSTENHOF, AND T. SØRENSEN, *Analyticity of the density of electronic wavefunctions*, Arkiv for Matematik, 42 (2004), pp. 87–106.

[55] ———, *Sharp regularity results for coulombic many-electron wave functions*, Communications in Mathematical Physics, 255 (2005), pp. 183–227.

Bibliography

[56] S. FOURNAIS, T. SØRENSEN, M. HOFFMANN-OSTENHOF, AND T. HOFFMANN-OSTENHOF, *Non-isotropic cusp conditions and regularity of the electron density of molecules at the nuclei*, Annales Henri Poincare, 8 (2007), pp. 731–748.

[57] M. FRAZIER, B. JAWERTH, AND G. WEISS, *Littlewood-Paley Theory and the Study of Function Spaces*, vol. 79 of CBMS, American Mathematical Society, 1991.

[58] G. FRIESECKE, *The multiconfiguration equations for atoms and molecules: Charge quantization and existence of solutions*, Archive for Rational Mechanics and Analysis, 169 (2003), pp. 35–71.

[59] M. GALASSI, J. DAVIES, J. THEILER, B. GOUGH, G. JUNGMAN, P. ALKEN, M. BOOTH, AND F. ROSSI, *GNU Scientific Library Reference Manual*, Network Theory Limited, 3 ed., 2009.

[60] J. GARCKE AND M. GRIEBEL, *On the computation of the eigenproblems of hydrogen and helium in strong magnetic and electric fields with the sparse grid combination technique*, Journal of Computational Physics, 165 (2000), pp. 694–716.

[61] W. GAUTSCHI, *Efficient computation of complex error function*, SIAM Journal on Numerical Analysis, 7 (1970), pp. 187–198.

[62] T. GERSTNER AND M. GRIEBEL, *Dimension–adaptive tensor–product quadrature*, Computing, 71 (2003), pp. 65–87.

[63] M. GRIEBEL, *Adaptive sparse grid multilevel methods for elliptic PDEs based on finite differences*, Computing, 61 (1998), pp. 151–179.

[64] ———, *Sparse grids and related approximation schemes for higher dimensional problems*, in Foundations of Computational Mathematics (FoCM05), Santander, L. Pardo, A. Pinkus, E. Suli, and M. Todd, eds., Cambridge University Press, 2006, pp. 106–161.

[65] M. GRIEBEL AND J. HAMAEKERS, *Molecular dynamics simulations of the elastic moduli of polymer-carbon nanotube composites*, Computer Methods in Applied Mechanics and Engineering, 193 (2004), pp. 1773–1788.

[66] ———, *Molecular dynamics simulations of the mechanical properties of polyethylene-carbon nanotube composites*, in Handbook of Theoretical and Computational Nanotechnology, M. Rieth and W. Schommers, eds., vol. 9, American Scientific Publishers, 2006, ch. 8, pp. 409–454.

[67] ———, *A wavelet based sparse grid method for the electronic Schrödinger equation*, in Proceedings of the International Congress of Mathematicians, M. Sanz-Solé, J. Soria, J. Varona, and J. Verdera, eds., vol. III, Madrid, Spain, August 22–30 2006, European Mathematical Society, pp. 1473–1506. Also as INS Preprint No. 0603, Institut für Numerische Simulation, Universität Bonn.

[68] ———, *Molecular dynamics, simulations of boron-nitride nanotubes embedded in amorphous Si-B-N*, Computational Materials Science, 39 (2007), pp. 502–517.

Bibliography

[69] ———, *Sparse grids for the Schrödinger equation*, Mathematical Modelling and Numerical Analysis, 41 (2007), pp. 215–247.

[70] M. GRIEBEL, J. HAMAEKERS, AND F. HEBER, *A molecular dynamics study on the impact of defects and functionalization on the Young modulus of boron-nitride nanotubes*, Computational Materials Science, 45 (2009), pp. 1097–1103.

[71] ———, *BOSSANOVA: A bond order dissection approach for efficient electronic structure calculations*, INS Preprint 0903, Institut für Numerische Simulation, Universität Bonn, 2009.

[72] M. GRIEBEL AND S. KNAPEK, *Optimized tensor-product approximation spaces*, Constructive Approximation, 16 (2000), pp. 525–540.

[73] ———, *Optimized general sparse grid approximation spaces for operator equations*, Mathematics of Computation, 78 (2009), pp. 2223–2257.

[74] M. GRIEBEL, S. KNAPEK, AND G. ZUMBUSCH, *Numerical Simulation in Molecular Dynamics*, Springer, 2007.

[75] M. GRIEBEL AND P. OSWALD, *Tensor product type subspace splitting and multilevel iterative methods for anisotropic problems*, Advances in Computational Mathematics, 4 (1995), pp. 171–206.

[76] M. GRIEBEL, P. OSWALD, AND T. SCHIEKOFER, *Sparse grids for boundary integral equations*, Numerische Mathematik, 83 (1999), pp. 279–312.

[77] M. GRIEBEL, M. SCHNEIDER, AND C. ZENGER, *A combination technique for the solution of sparse grid problems*, in Iterative Methods in Linear Algebra, P. de Groen and R. Beauwens, eds., IMACS, Elsevier, North Holland, 1992, pp. 263–281.

[78] W. GROPP, E. LUSK, AND A. SKJELLUM, *Using MPI: Portable Parallel Programming with the Message Passing Interface*, MIT Press, 2 ed., 1999.

[79] W. HACKBUSCH, *The efficient computation of certain determinants arising in the treatment of Schrödinger's equations*, Computing, 67 (2001), pp. 35–56.

[80] W. HACKBUSCH AND B. KHOROMSKIJ, *Low-rank Kronecker-product approximation to multi-dimensional nonlocal operators. Part I. Separable approximation of multi-variate functions*, Computing, 76 (2006), pp. 177–202.

[81] ———, *Tensor-product approximation to operators and functions in high dimensions*, Journal of Complexity, 23 (2007), pp. 697–714.

[82] D. HAROSKE AND H. TRIEBEL, *Distributions, Sobolev Spaces, Elliptic Equations*, European Mathematical Society, 2007.

[83] R. HARRISON, G. FANN, T. YANAI, Z. GAN, AND G. BEYLKIN, *Multiresolution quantum chemistry: Basic theory and initial applications*, Journal of Chemical Physics, 121 (2004), pp. 11587–11598.

Bibliography

[84] M. HECKERT, M. KALLAY, D. TEW, W. KLOPPER, AND J. GAUSS, *Basis-set extrapolation techniques for the accurate calculation of molecular equilibrium geometries using coupled-cluster theory*, Journal of Chemical Physics, 125 (2006), p. 044108.

[85] T. HELGAKER, P. JØRGENSEN, AND J. OLSEN, *Molecular electronic structure theory*, John Wiley and Sons, 2001.

[86] T. HELGAKER, W. KLOPPER, AND D. TEW, *Quantitative quantum chemistry*, Molecular Physics, 116 (2008), pp. 2107–2143.

[87] E. HERNÁNDEZ AND G. WEISS, *A First Course on Wavelets*, CRC Press, 1996.

[88] V. HERNANDEZ, J. ROMAN, E. ROMERO, A. TOMAS, AND V. VIDAL, *SLEPc users manual*, Tech. Rep. DSIC-II/24/02 - Revision 2.3.3, D. Sistemas Informáticos y Computación, Universidad Politécnica de Valencia, 2007.

[89] V. HERNANDEZ, J. ROMAN, A. TOMAS, AND V. VIDAL, *Krylov-Schur methods in SLEPc*, Tech. Rep. STR-7, Universidad Politécnica de Valencia, 2007. Available at http://www.grycap.upv.es/slepc.

[90] ——, *A survey of software for sparse eigenvalue problems*, Tech. Rep. STR-6, Universidad Politécnica de Valencia, 2007. Available at http://www.grycap.upv.es/slepc.

[91] V. HERNANDEZ, J. ROMAN, AND V. VIDAL, *SLEPc: A scalable and flexible toolkit for the solution of eigenvalue problems*, ACM Transactions on Mathematical Software, 31 (2005), pp. 351–362.

[92] R. HILL, *Completeness of Gaussian orbital and geminal basis sets for linear molecules in L^2 and in the first and second Sobolev spaces*, International Journal of Quantum Chemistry, 68 (1998), pp. 357–384.

[93] R. HOCHMUTH, S. KNAPEK, AND G. ZUMBUSCH, *Tensor products of Sobolev spaces and applications*. Technical Report 685, SFB 256, Universität Bonn, 2000.

[94] M. HOFFMANN-OSTENHOF, T. HOFFMANN-OSTENHOF, AND H. STREMNITZER, *Electronic wave-functions near coalescence points*, Physical Review Letters, 68 (1992), pp. 3857–3860.

[95] ——, *Local properties of Coulombic wave-functions*, Communications in Mathematical Physics, 163 (1994), pp. 185–215.

[96] M. HOLTZ, *Sparse Grid Quadrature in High Dimensions with Applications in Finance and Insurance*, Dissertation, Institut für Numerische Simulation, Universität Bonn, November 2008.

[97] O. HOLTZ AND A. RON, *Approximation orders of shift-invariant subspaces of $W_2^s(\mathbb{R}^d)$*, Journal of Approximation Theory, 132 (2005), pp. 97–148.

[98] W. HUNZIKER AND I. SIGAL, *The quantum N-body problem*, Journal of Mathematical Physics, 41 (2000), pp. 3448–3510.

[99] K. ISHIDA, *Accurate and fast algorithm of the molecular incomplete gamma function with a complex argument*, Journal of Computational Chemistry, 25 (2004), pp. 739–748.

[100] M. JOHNSON, *Approximation in $L^p(\mathbb{R}^d)$ from spaces spanned by the perturbed integer translates of a radial function*, Journal of Approximation Theory, 107 (2000), pp. 163–203.

[101] M. JONES, G. ORTIZ, AND D. CEPERLEY, *Hartree-Fock studies of atoms in strong magnetic fields*, Physical Review A, 54 (1996), pp. 219–231.

[102] M. JONES AND P. PLASSMANN, *BlockSolve95 users manual: Scalable library software for the parallel solution of sparse linear systems*, Tech. Rep. ANL-95/48, Argonne National Laboratory, 1995.

[103] N. KAIBLINGER AND W. MADYCH, *Orthonormal sampling functions*, Applied and Computational Harmonic Analysis, 21 (2006), pp. 404–412.

[104] T. KATO, *On the eigenfunctions of many-particle systems in quantum mechanics*, Communications on Pure and Applied Mathematics, 10 (1957), pp. 151–177.

[105] W. KLOPPER, F. MANBY, S. TEN-NO, AND E. VALEEV, *R12 methods in explicitly correlated molecular electronic structure theory*, International Reviews in Physical Chemistry, 25 (2006), pp. 427–468.

[106] S. KNAPEK, *Approximation und Kompression mit Tensorprodukt-Multiskalenräumen*, Dissertation, Institut für Angewandte Mathematik, Universität Bonn, April 2000.

[107] S. KNAPEK, *Hyperbolic cross approximation of integral operators with smooth kernel*. Technical Report 665, SFB 256, Universität Bonn, 2000.

[108] D. KNUTH, *Sorting and searching*, vol. 3 of The art of computer programming, Addison-Wesley, 1973.

[109] A. KNYAZEV, *Toward the optimal preconditioned eigensolver: Locally optimal block preconditioned conjugate gradient method*, SIAM Journal on Scientific Computing, 23 (2001), pp. 517–541.

[110] T. KOLDA, *Orthogonal tensor decompositions*, SIAM Journal on Matrix Analysis and Applications, 23 (2001), pp. 243–255.

[111] W. KOLOS AND L. WOLNIEWI, *Improved theoretical ground-state energy of hydrogen molecule*, Journal of Chemical Physics, 49 (1968), pp. 404–410.

[112] F. KUO, I. SLOAN, G. WASILKOWSKI, AND H. WOŹNIAKOWSKI, *On decompositions of multivariate functions*, Applied Mathematics Report AMR08/05, University of New South Wales, 2008.

[113] W. KUTZELNIGG, *Convergence of expansions in a Gaussian basis*, in Strategies and Applications in Quantum Chemistry, M. Defranceschi and Y. Ellinger, eds., vol. 14 of Topics in Molecular Organization and Engineering, Kluwer Academic Publishers, 1996, pp. 79–101.

[114] W. KUTZELNIGG AND W. KLOPPER, *Wave-functions with terms linear in the interelectronic coordinates to take care of the correlation cusp. I. General-theory*, Journal of Chemical Physics, 94 (1991), pp. 1985–2001.

[115] W. KUTZELNIGG AND D. MUKHERJEE, *Minimal parametrization of an n-electron state*, Physical Review A, 71 (2005), p. 022502.

[116] G. KYRIAZIS, *Wavelet-type decompositions and approximations from shift-invariant spaces*, Journal of Approximation Theory, 88 (1997), pp. 257–271.

[117] ———, *Decomposition systems for function spaces*, Studia Mathematica, 157 (2003), pp. 133–169.

[118] G. KYRIAZIS AND P. PETRUSHEV, *New bases for Triebel-Lizorkin and Besov spaces*, Transactions of the American Mathematical Society, 354 (2002), pp. 749–776.

[119] ———, *On the construction of frames for Triebel-Lizorkin and Besov spaces*, Proceedings of the American Mathematical Society, 134 (2006), pp. 1759–1770.

[120] C. LE BRIS, *Computational chemistry from the perspective of numerical analysis*, Acta Numerica, 14 (2005), pp. 363–444.

[121] C. LE BRIS AND P. LIONS, *From atoms to crystals: A mathematical journey*, Bulletin of the American Mathematical Society, 42 (2005), pp. 291–363.

[122] R. LEHOUCQ, D. SORENSEN, AND C. YANG, *ARPACK Users' Guide: Solution of Large-Scale Eigenvalue Problems with Implicitly Restarted Arnoldi Methods*, SIAM, 1998.

[123] M. LEWIN, *Solutions of the multiconfiguration equations in quantum chemistry*, Archive for Rational Mechanics and Analysis, 171 (2004), pp. 83–114.

[124] E. LIEB, *Bound on the maximum negative ionization of atoms and molecules*, Physical Review A, 29 (1984), pp. 3018–3028.

[125] ———, *The stability of matter - from atoms to stars*, Bulletin of the American Mathematical Society, 22 (1990), pp. 1–49.

[126] P.-O. LÖWDIN, *Quantum theory of many-particle systems. I. Physical interpretations by means of density matrices, natural spin-orbitals, and convergence problems in the method of configurational interaction*, Physical Review, 97 (1955), pp. 1474–1489.

[127] A. LÜCHOW AND J. ANDERSON, *First-row hydrides: Dissociation and ground state energies using quantum Monte Carlo*, Journal of Chemical Physics, 105 (1996), pp. 7573–7578.

[128] W. LUI AND Y. CHEN, *Wavelet and multiple scale reproducing kernel methods*, International Journal for Numerical Methods in Fluids, 21 (1995), pp. 901–931.

[129] H. LUO, D. KOLB, H. FLAD, W. HACKBUSCH, AND T. KOPRUCKI, *Wavelet approximation of correlated wave functions. II. Hyperbolic wavelets and adaptive approximation schemes*, Journal of Chemical Physics, 117 (2002), pp. 3625–3638.

[130] S. MALLAT, *A Wavelet Tour of Signal Processing*, Academic Press, 2 ed., 1999.

[131] K. MASCHHOFF AND D. SORENSEN, *P_ARPACK: An efficient portable large scale eigenvalue package for distributed memory parallel architectures*, in Applied Parallel Computing in Industrial Problems and Optimization, J. Wasniewski, J. Dongarra, K. Madsen, and D. Olesen, eds., Springer, 1996.

[132] R. MATHAR, *Numerical representations of the incomplete gamma function of complex-valued argument*, Numerical Algorithms, 36 (2004), pp. 247–264.

[133] V. MAZ'YA AND G. SCHMIDT, *On approximate approximations using gaussian kernels*, Ima Journal of Numerical Analysis, 16 (1996), pp. 13–29.

[134] ———, *Approximate wavelets and the approximation of pseudodifferential operators*, Applied and Computational Harmonic Analysis, 6 (1999), pp. 287–313.

[135] ———, *Approximate Approximation*, Mathematical Surveys and Monographs, American Mathematical Society, 2007.

[136] L. MCMURCHIE AND E. DAVIDSON, *One- and two-electron integrals over cartesian Gaussian functions*, Journal of Computational Physics, 26 (1978), pp. 218–231.

[137] A. MESSIAH, *Quantum Mechanics: Two Volumes Bound as One*, Courier Dover Publications, 1999.

[138] S. J. Y. MEYER AND R. RYAN, *Wavelets: Tools for Science & Technology*, SIAM, Philadelphia, 2001.

[139] Y. MEYER, *Wavelets and operators*, vol. 37 of Cambridge Studies in Advanced Mathematics, Cambridge University Press, 1992.

[140] H. NAKASHIMA, Y. HIJIKATA, AND H. NAKATSUJI, *Solving the electron and electron-nuclear Schrödinger equations for the excited states of helium atom with the free iterative-complement-interaction method*, Journal of Chemical Physics, 128 (2008), p. 154108.

[141] H. NAKATSUJI, *Scaled Schrödinger equation and the exact wave function*, Physical Review Letters, 93 (2004), p. 030403.

[142] ———, *General method of solving the Schrödinger equation of atoms and molecules*, Physical Review A, 72 (2005), p. 062110.

[143] A. NEELOV AND S. GOEDECKER, *An efficient numerical quadrature for the calculation of the potential energy of wavefunctions expressed in the daubechies wavelet basis*, Journal of Computational Physics, 217 (2006), pp. 312–339.

[144] J. NELDER AND R. MEAD, *A simplex-method for function minimization*, Computer Journal, 7 (1965), pp. 308–313.

[145] M. NIGHTINGALE AND C. UMRIGAR, eds., *Quantum Monte Carlo Methods in Physics and Chemistry*, vol. 525 of NATO Science Series C, Springer, 1999.

[146] P. NITSCHE, *Best N term approximation spaces for tensor product wavelet bases*, Constructive Approximation, 24 (2006), pp. 49–70.

[147] M. NOOIJEN AND R. BARTLETT, *Elimination of Coulombic infinities through transformation of the Hamiltonian*, Journal of Chemical Physics, 109 (1998), pp. 8232–8240.

[148] T. OTTMANN AND P. WIDMAYER, *Algorithmen und Datenstrukturen*, BI Wissenschaftsverlag, 2 ed., 1993.

[149] R. PACK AND W. BYERS BROWN, *Cusp conditions for molecular wavefunctions*, Journal of Chemical Physics, 45 (1966), pp. 556–559.

[150] R. PARR AND W. YANG, *Density-Functional Theory of Atoms and Molecules*, Oxford University Press, 1994.

[151] A. PERSSON, *Bounds for the discrete part of the spectrum of a semi-bounded Schrödinger operator*, Mathematica Scandinavica, 8 (1960), pp. 143–153.

[152] I. PETSALAKIS, G. THEODORAKOPOULOS, C. NICOLAIDES, R. BUENKER, AND S. PEYERIMHOFF, *Nonorthonormal CI for molecular excited states. I. The sudden polarization effect in 90° twisted ethylene*, The Journal of Chemical Physics, 81 (1984), pp. 3161–3167.

[153] G. POPPE AND C. WIJERS, *Evaluation of the complex error function*, ACM Transactions on Mathematical Software, 16 (1990), pp. 47–47.

[154] V. POSPELOV, *On the theory of singular expansion in a tensor product of Hilbert-spaces*, Russian Academy of Sciences Sbornik Mathematics, 82 (1995), pp. 357–364.

[155] F. PROSSER AND S. HAGSTROM, *On the rapid computation of matrix elements*, International Journal of Quantum Chemistry, 2 (1968), pp. 89–99.

[156] H. RABITZ AND Ö. ALIŞ, *General foundations of high-dimensional model representations*, Journal of Mathematical Chemistry, 25 (1999), pp. 197–233.

[157] V. RASSOLOV AND D. CHIPMAN, *Behavior of electronic wave functions near cusps*, Journal of Chemical Physics, 104 (1996), pp. 9908–9912.

[158] I. RØEGGEN, *Extended geminal models*, in Correlation and Localization, P. Surján, ed., vol. 203 of Topics in Current Chemistry, Springer, 1999, pp. 89–103.

Bibliography

[159] U. SCHERZ, *Quantenmechanik: Eine Einführung mit Anwendungen auf Atome, Moleküle und Festkörper*, Teubner, 1999.

[160] H. SCHMEISSER AND H. TRIEBEL, *Fourier Analysis and Functions Spaces*, John Wiley, 1987.

[161] R. SCHNEIDER, *Analysis of the projected coupled cluster method in electronic structure calculation*, ePrint 2008-007, Institut für Mathematik, Technische Universität Berlin, Germany, 2008.

[162] W. SICKEL AND T. ULLRICH, *Tensor products of Sobolev-Besov spaces and applications to approximation from the hyperbolic cross*, Journal of Approximation Theory, 161 (2009), pp. 748–786.

[163] B. SIMON, *Schrödinger operators in the twentieth century*, Journal of Mathematical Physics, 41 (2000), pp. 3523–3555.

[164] O. SINANOĞLU, *A method for the analysis of many-electron wave functions*, Reviews of Modern Physics, 35 (1963), pp. 517–519.

[165] K. SINGER, *The use of Gaussian (exponential quadratic) wave functions in molecular problems. I. General formulae for the evaluation of integrals*, Proceedings of the Royal Society of London Series A: Mathematical and Physical Sciences, 258 (1960), pp. 412–420.

[166] J. SLATER, *The theory of complex spectra*, Physical Review, 34 (1929), pp. 1293–1322.

[167] I. SLOAN, X. WANG, AND H. WOŹNIAKOWSKI, *Finite-order weights imply tractability of multivariate integration*, Journal of Complexity, 20 (2004), pp. 46–74.

[168] G. STEWART, *A Krylov–Schur algorithm for large eigenproblems*, SIAM Journal on Matrix Analysis and Applications, 23 (2001), pp. 601–614.

[169] P. SURJÁN, *An introduction to the theory of geminals*, in Correlation and Localization, P. Surján, ed., vol. 203 of Topics in Current Chemistry, Springer, 1999, pp. 63–88.

[170] A. SZABO AND N. OSTLUND, *Modern Quantum Chemistry, Introduction to Advanced Electronic Structure Theory*, Dover Publications, 1996.

[171] M. TACHIKAWA AND M. SHIGA, *Evaluation of atomic integrals for hybrid gaussian type and plane-wave basis functions via the McMurchie-Davidson recursion formula*, Physical Review E, 64 (2001), p. 056706.

[172] G. TASI AND A. CSÁSZÁR, *Hartree-Fock-limit energies and structures with a few dozen distributed Gaussians*, Chemical Physics Letters, 438 (2007), pp. 139–143.

[173] S. TEN-NO, *Initiation of explicitly correlated Slater-type geminal theory*, Chemical Physics Letters, 398 (2004), pp. 56–61.

Bibliography

[174] D. TEW, *Second order coalescence conditions of molecular wave functions*, Journal of Chemical Physics, 129 (2008), p. 014104.

[175] D. TEW, W. KLOPPER, AND T. HELGAKER, *Electron correlation: The many-body problem at the heart of chemistry*, Journal of Computational Chemistry, 28 (2007), pp. 1307–1320.

[176] T. TORSTI, T. EIROLA, J. E. T., P. HAKALA, P. HAVU, V. HAVU, T. HOYNALANMAA, J. IGNATIUS, M. LYLY, I. MAKKONEN, T. RANTALA, J. RUOKOLAINEN, K. RUOTSALAINEN, E. RASANEN, H. SAARIKOSKI, AND M. PUSKA, *Three real-space discretization techniques in electronic structure calculations*, physica status solidi (b) - basic solid state physics, 243 (2006), pp. 1016–1053.

[177] J. TOULOUSE AND C. UMRIGAR, *Full optimization of Jastrow-Slater wave functions with application to the first-row atoms and homonuclear diatomic molecules*, Journal of Chemical Physics, 128 (2008), p. 174101.

[178] H. TRIEBEL, *Theory of Function Spaces III*, Monographs in Mathematics, Birkhäuser Verlag, 2006.

[179] T. ULLRICH, *Function spaces with dominating mixed smoothness, characterization by differences*, Jenaer Schriften zur Mathematik und Informatik, Math/Inf/05/06 (2006), pp. 1–50.

[180] R. VAN DE GEIJN, P. ALPATOU, G. BAKER, C. EDWARDS, J. GUNNELS, G. MORROW, AND J. OVERFELT, *Using PLAPACK: Parallel Linear Algebra Package*, MIT Press, 1997.

[181] J. VERBEEK AND J. VAN LENTHE, *On the evaluation of non-orthogonal matrix elements*, Journal of Molecular Structure: THEOCHEM, 229 (1991), pp. 115–137.

[182] J. VYBIRAL, *Function spaces with dominating mixed smoothness*, Dissertation, Fakultät für Mathematik und Informatik, Friedrich-Schiller-Universität Jena, 2005.

[183] D. WALNUT, *An introduction to wavelet analysis*, Birkhäuser, 2002.

[184] G. WASILKOWSKI AND H. WOŹNIAKOWSKI, *Finite-order weights imply tractability of linear multivariate problems*, Journal of Approximation Theory, 130 (2004), pp. 57–77.

[185] ———, *Polynomial-time algorithms for multivariate linear problems with finite-order weights: Average case setting*, Foundations of Computational Mathematics, 9 (2009), pp. 105–132.

[186] J. WEIDMANN, *Linear Operators in Hilbert Spaces*, Graduate Texts in Mathematics, Springer, 1980.

[187] B. WELLS AND S. WILSON, *On the accuracy of the algebraic-approximation in molecular electronic-structure calculations: I. Calculations for H_2^+, HeH^{2+}, H_2 and HeH^+ using basis-sets of atom-centered Gaussian-type functions*, Journal of Physics B-Atomic Molecular and Optical Physics, 22 (1989), pp. 1285–1295.

Bibliography

[188] A. WERSCHULZ AND H. WOŹNIAKOWSKI, *Tractability of quasilinear problems I: General results*, Journal of Approximation Theory, 145 (2007), pp. 266–285.

[189] D. WOON AND T. DUNNING, *Gaussian-basis sets for use in correlated molecular calculations. 4. Calculation of static electrical response properties*, Journal of Chemical Physics, 100 (1994), pp. 2975–2988.

[190] M. YAMADA AND K. OHKITANI, *An identification of energy cascade in turbulence by orthonormal wavelet analysis*, Progress of Theoretical Physics, 86 (1991), pp. 799–815.

[191] K. YAMAGUCHI AND T. MUKOYAMA, *Wavelet representation for the solution of radial Schrödinger equation*, Journal of Physics B - Atomic Molecular and Optical Physics, 29 (1996), pp. 4059–4071.

[192] H. YSERENTANT, *On the electronic Schrödinger equation*. Report 191, SFB 328, Universität Tübingen, 2003.

[193] ———, *On the regularity of the electronic Schrödinger equation in Hilbert spaces of mixed derivatives*, Numerische Mathematik, 98 (2004), pp. 731–759.

[194] ———, *Sparse grid spaces for the numerical solution of the electronic Schrödinger equation*, Numerische Mathematik, 101 (2005), pp. 381–389.

[195] ———, *The hyperbolic cross space approximation of electronic wavefunctions*, Numerische Mathematik, 105 (2007), pp. 659–690.

[196] ———, *Regularity properties of wavefunctions and the complexity of the quantum-mechanical N-body problem*. Preprint, November 2007. http://www.math.tu-berlin.de/~yserenta/Publikationen/papers/qm8.pdf.

[197] F. ZHANG, ed., *The Schur Complement and Its Applications*, vol. 4 of Numerical Methods and Algorithms, Springer, 2005.

I want morebooks!

Buy your books fast and straightforward online - at one of the world's fastest growing online book stores! Environmentally sound due to Print-on-Demand technologies.

Buy your books online at
www.get-morebooks.com

Kaufen Sie Ihre Bücher schnell und unkompliziert online – auf einer der am schnellsten wachsenden Buchhandelsplattformen weltweit!
Dank Print-On-Demand umwelt- und ressourcenschonend produziert.

Bücher schneller online kaufen
www.morebooks.de

OmniScriptum Marketing DEU GmbH
Heinrich-Böcking-Str. 6-8
D - 66121 Saarbrücken
Telefax: +49 681 93 81 567-9

info@omniscriptum.com
www.omniscriptum.com

Printed by Books on Demand GmbH, Norderstedt / Germany